MECHANICAL ENGINEERING THEORY AND APPLICATIONS

PROGRESS IN ANALYSIS OF FUNCTIONALLY GRADED STRUCTURES

MECHANICAL ENGINEERING THEORY AND APPLICATIONS

Additional books in this series can be found on Nova's website under the Series tab.

Additional E-books in this series can be found on Nova's website under the E-books tab.

MECHANICAL ENGINEERING THEORY AND APPLICATIONS

PROGRESS IN ANALYSIS OF FUNCTIONALLY GRADED STRUCTURES

FARZAD EBRAHIMI,
HOSSEIN SEPIANI
AND
ALI GHORBANPOUR ARANI

Nova Science Publishers, Inc.
New York

For permission to use material from this book please contact us:
Telephone 631-231-7269; Fax 631-231-8175
Web Site: http://www.novapublishers.com

LIBRARY OF CONGRESS CATALOGING-IN-PUBLICATION DATA

Ibrahimi, Farzad.
Progress in analysis of functionally graded structures / authors: Farzad Ebrahimi, Hosein Ali Sepiani, Ali Ghorbanpour Arani.
p. cm.
Includes index.
ISBN 978-1-61209-837-1 (softcover)
1. Shells (Engineering)--Materials. 2. Structural analysis (Engineering) 3. Functionally gradient materials. I. Sepiani, Hosein Ali. II. Arani, Ali Ghorbanpour Arani. III. Title.
TA660.S5I37 2011
624.1'7762--dc22

2011004609

Published by Nova Science Publishers, Inc. † New York

CONTENTS

PREFACE

Functionally graded materials (FGMs) are multiphase composites with continuously varying volume fractions and, consequently, thermo- mechanical properties. The concept of grading the thermo-mechanical properties of materials provides an important tool to design new materials for certain specific functions. To take full advantage of this new tool research is needed not only for developing efficient processing methods and material characterization techniques but also for carrying out basic mechanics studies to the safety and durability of the FGM components.

FGMs were initially designed as thermal barrier materials for aerospace structural applications and fusion reactors. They are now developed for general use as structural components in extremely high-temperature environments. Low thermal conductivity, low coefficient of thermal expansion and core ductility have enabled the FGM materials to withstand higher temperature gradients for a given heat flux. Thus, the use of FGM may become an important issue for developing advanced structures and it is therefore of prime importance to express newly developed analytical solution methods in analysis of FGM structures.

Consequently, the aim of this book is to present derivations of the basic equations of mechanics in invariant (vector and tensor) form and specializations of the governing equations of Thermoelastic, Magnetothermoelastic, vibration and buckling analysis to both thin and thick shells and spheres made of functionally graded materials in the form of both single and double layer as well as the smart ones; numerous illustrative examples; and chapter-end summaries. This book consists of seven chapters: In Chapter 1, the vibration and buckling analysis of thin and thick cylindrical shells made of functionally graded materials are introduced considering the effects of transverse shear and rotary inertias while the dynamic stability analysis of both thin and thick cylindrical FGM shells is presented in is Chapter 2. Chapter 3 furnishes a detailed treatment of the vibration and buckling analysis of two-layered FG shells while it is extended to the elastic stability analysis of two-layered functionally graded cylindrical shells in chapter 4. Finally Magnetothermoelastic stress transient response of functionally graded thick hollow sphere subjected to magnetic and thermoelastic fields are introduced in chapters 5, 6.

All of the solutions presented in these chapters are the results of recent investigations conducted by the authors and their collaborators. The results presented herein may be treated as a benchmark for checking the validity and accuracy of other numerical solutions. Despite a number of existing texts on the theory and analysis of plates and/or shells, there is not a single book that is devoted entirely to the analysis of inhomogeneous isotropic and functionally graded shells and spheres. It is hoped that this book will fill the gap to some extent and be used as a valuable reference source for postgraduate students, engineers, scientists, and applied mathematicians in this field and the authors believe that the reader who takes time to study this book will find ample reward.

Chapter 1

VIBRATION AND BUCKLING ANALYSIS OF FGM CYLINDRICAL SHELLS UNDER COMBINED STATIC AND PERIODIC AXIAL FORCES

ABSTRACT

In this study, a formulation for the free vibration and buckling of cylindrical shells made of functionally graded material (FGM) subjected to combined static and periodic axial loadings are presented. The properties are temperature dependent and graded in the thickness direction according to a volume fraction power law distribution. The analysis is based on two different methods of first-order shear deformation theory (FSDT) considering the transverse shear strains and the rotary inertias and the classical shell theory (CST). The results obtained show that the effect of transverse shear and rotary inertias on vibration and buckling of functionally graded cylindrical shells is dependent on the material composition, the temperature environment, the amplitude of static load, the deformation mode, and the shell geometry parameters.

1. INTRODUCTION

Functionally graded materials (FGMs) are being increasingly considered in various applications to maximize strengths and integrities of many engineering structures. FGMs have received considerable attention in many engineering applications since they were first reported in 1984 in Japan (see [1]). FGMs are composite materials, microscopically inhomogeneous, in which the mechanical properties vary smoothly and continuously from one surface to the other. This is achieved by gradually varying the volume fraction of the constituent materials. FGMs were initially designed as thermal barrier materials for aerospace structures and fusion reactors and now they are also considered as potential structural materials for the future high-speed spacecrafts. Formulation and theoretical analysis of the FGM plates and shells were presented by Reddy and chin [2, 3], Arciniega and Reddy [4] and Praveen et al. [5]. Shell structure made up of this composite material (FGM) is also one of the basic structural elements used in many engineering structures.

Despite the evident importance in practical applications, investigations on the static and dynamic characteristics of FGM shell structures are still limited in number. Among those available, Loy et al. [6] investigated the free vibration of simply supported functionally graded (FG) cylindrical shells, which was later extended by Pradhan et al. [7] to cylindrical shells under various end supporting conditions. Gong et al.[8] presented elastic response analysis of simply supported FGM cylindrical shells under low-velocity impact. Ng et al. [9] studied dynamic instability of simply supported FGM cylindrical shells, a normal-mode expansion and Bolotin method were used to determine the boundaries of the unstable regions. In all the above studies, theoretical formulations were all based on classical shell theory, i.e., neglecting the effect of transverse shear strains. By using the first-order shear deformation shell theory, Jafari et al. [10] investigated the vibration and buckling of cross-ply laminated circular cylindrical shells. Bhangale and Ganesan [11] solved the free vibration problem for simply supported non-homogeneous FG graded magneto-electro-elastic cylindrical shells using finite element method. Using Love's shell theory and the Galerkin method, Ravikiran and Ganesan [12] investigated the buckling and free vibration of FGM cylindrical shells using the finite element models of shell.

This chapter studies the effect of transverse shear and rotary inertias on free vibration and buckling of the FG cylindrical shells subjected to combined static and periodic axial forces, through a comparison of results obtained by using two different methods such as the first-order shear deformation theory (FSDT) considering the transverse shear strains and the rotary inertias and the classical shell theory (CST). Material properties of functionally graded cylindrical shells are considered as temperature dependent and graded in the thickness direction according to a power-law distribution in terms of the volume fractions of the constituents. The results obtained show that the effect of transverse shear and rotary inertias on free vibration and buckling of FG cylindrical shells subjected to combined static and periodic axial forces is dependent on the material composition, the temperature environment, the amplitude of static load, the deformation mode, and the shell geometry parameters. It is found that the effect of transverse shear and rotary inertias on free vibration and buckling of FG cylindrical shells subjected to combined static and periodic axial forces is not neglected in some cases. The new features of the effect of transverse shear and rotary inertias on free vibration and buckling of FG cylindrical shells and some meaningful results in this chapter are helpful for the application and the design of nuclear reactors, space planes and chemical plants, in which FG cylindrical shells act as basic elements.

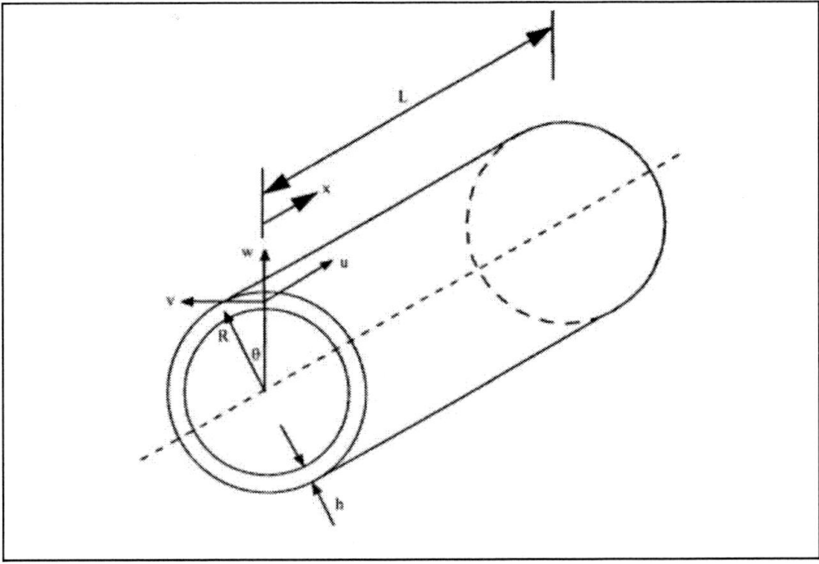

Figure 1. Coordinate system of the FGM cylindrical shell.

2. Theoretical Formulations

An functionally graded materials cylindrical shell with mean radius of R, thickness h, and the length L is shown in Figure 1. The displacement components in the x, θ and z direction are denoted by u, v and w respectively. The pulsating axial load is given by

Na=N0+Nd cos Pt (1)

where P is the frequency of excitation in radians per unit time.

The material properties of FGM cylindrical shells with both temperature dependent and position dependent are accurately modeled, by using a simple rule of mixtures for the stiffness parameters coupled with the temperature dependent properties of the constituents. The volume fraction is described by a spatial function as follows

$$V(z) = (z/h + 1/2)^{\Phi} \quad , \quad (0 \le \Phi \le \infty) \tag{2}$$

where Φ expresses the volume fraction exponent. The combination of these functions gives rise to the effective properties of functionally graded materials. An FGM cylindrical shell that is metal rich at the inner surface and ceramic rich at the outer surface is defined as Type A. The corresponding effective material properties are expressed as

$$F_{eff}(T,z) = F_c(T)V(z) + F_m(T)(1 - V(z)) \tag{3}$$

where F_{eff} is the effective material property of the FGM cylindrical shell, including the effective elastic modulus, effective mass density and effective Poisson's ratio. F_c and F_m are the temperature dependent properties of the ceramic and metal, respectively . On the other hand, an FGM cylindrical shell that is ceramic rich at the inner surface and metal rich at the outer surface is defined as Type B, whose effective material properties are given by

$$F_{eff}(T,z) = F_m(T)V(z) + F_c(T)(1 - V(z)) \tag{4}$$

Based on the first-order shear deformation theory (FSDT), the equations of motion for an FGM cylindrical shell under axially dynamic load are as follows [10, 13]

$$\frac{\partial N_1}{\partial x} + \frac{1}{R}\frac{\partial N_6}{\partial \theta} = I_1 u'' + I_2 \phi_1'' \tag{5}$$

$$\frac{\partial N_6}{\partial x}+\frac{1}{R}\frac{\partial N_1}{\partial \theta}+\frac{1}{R}Q_2+N_a\frac{\partial^2 v}{\partial x^2}=I_1 v''+I_2 \phi_2'' \tag{6}$$

$$\frac{\partial Q_1}{\partial x}+\frac{1}{R}\frac{\partial Q_2}{\partial \theta}-\frac{N_2}{R}+N_a\frac{\partial^2 w}{\partial x^2}=I_1 w'' \tag{7}$$

$$\frac{\partial M_1}{\partial x}+\frac{1}{R}\frac{\partial M_6}{\partial \theta}-Q_1=I_2 u''+I_3 \phi_1'' \tag{8}$$

$$\frac{\partial M_6}{\partial x}+\frac{1}{R}\frac{\partial M_2}{\partial \theta}-Q_2=I_2 v''+I_3 \phi_2'' \tag{9}$$

where ϕ_1 and ϕ_2 are the rotations of a normal to the reference surface, $I_i(i=1,2,3)$ is the mass inertia terms defined as

$$\left(I_1,I_2,I_3\right)=\int_{-\frac{h}{2}}^{\frac{h}{2}}\rho(z)(1,z,z^2)dz \tag{10}$$

and $\rho(z)$ is the effective mass density of functionally graded materials.

The stress resultants of FGM cylindrical shells are given by

$$\begin{pmatrix} N_1 \\ N_2 \\ N_6 \\ M_1 \\ M_2 \\ M_6 \end{pmatrix}=\begin{pmatrix} A_{11} & A_{12} & 0 & B_{11} & B_{12} & 0 \\ A_{21} & A_{22} & 0 & B_{21} & B_{22} & 0 \\ 0 & 0 & A_{66} & 0 & 0 & B_{66} \\ B_{11} & B_{12} & 0 & D_{11} & D_{12} & 0 \\ B_{21} & B_{22} & 0 & D_{21} & D_{22} & 0 \\ 0 & 0 & B_{66} & 0 & 0 & D_{66} \end{pmatrix}\begin{pmatrix} \varepsilon_1 \\ \varepsilon_2 \\ \varepsilon_6 \\ \kappa_1 \\ \kappa_2 \\ \kappa_6 \end{pmatrix}$$

$$\begin{pmatrix} Q_1 \\ Q_2 \end{pmatrix} == \begin{pmatrix} C_{44} & 0 \\ 0 & C_{55} \end{pmatrix} \begin{pmatrix} \varepsilon_5 \\ \varepsilon_4 \end{pmatrix} \tag{11}$$

where A_{ij}, B_{ij}, D_{ij} and C_{ij} are ,respectively, the extensional, coupling, bending, and shear stiffness, which are given by

$$\left(A_{ij}, B_{ij}, D_{ij}\right) = \int_{-\frac{h}{2}}^{\frac{h}{2}} Q_{ij}(1, z, z^2)dz\,, \quad (i = 1,2,6)$$

$$\begin{aligned}
Q_{11} &= Q_{22} = \frac{E_{eff}}{1-\nu_{eff}^{\ 2}} & C_{44} &= \int_{-\frac{h}{2}}^{\frac{h}{2}} Q_{44}dz \\
Q_{12} &= Q_{21} = \frac{\nu_{eff} E_{eff}}{A(1-\nu_{eff}^{\ 2})} & C_{55} &= \int_{-\frac{h}{2}}^{\frac{h}{2}} Q_{55}dz \\
Q_{44} &= \kappa \frac{E_{eff}}{2\left(1+\nu_{eff}\right)} & Q_{66} &= \frac{E_{eff}}{2A(1+\nu_{eff})} \\
Q_{55} &= Q_{44}/A & A &= 1+z/R
\end{aligned} \tag{12}$$

where E_{eff} and ν_{eff} are the effective elastic modulus and effective Poisson's ratio of FGM cylindrical shells, respectively which are defined according to Eq. (3). κ is the shear correction factor introduced by Reddy [2] and is equal to 5/6 . The strains are expressed as

$$\begin{aligned}
&\varepsilon_1 = \frac{\partial u}{\partial x} \ \varepsilon_2 = \frac{1}{R}\left(\frac{\partial v}{\partial \theta} + w\right) \varepsilon_6 = \frac{\partial v}{\partial x} + \frac{1}{R}\frac{\partial u}{\partial \theta} \ \varepsilon_4 = \phi_2 + \frac{1}{R}\frac{\partial w}{\partial \theta} \\
&\varepsilon_5 = \phi_1 + \frac{\partial w}{\partial x} \ \kappa_1 = \frac{\partial \phi_1}{\partial x} \ \kappa_2 = \frac{1}{R}\frac{\partial \phi_2}{\partial \theta} \ \kappa_6 = \frac{\partial \phi_2}{\partial x} + \frac{1}{R}\frac{\partial \phi_1}{\partial \theta}
\end{aligned} \tag{13}$$

Utilizing Eqs.(5)-(9), (11) and (13), the equations of motion can be expressed in terms of generalized displacement $(u, v, w, \phi_1, \phi_2)$ as follows

$$L_1(u, v, w, \phi_1, \phi_2) = I_1 u'' + I_2 \phi_1'' \tag{14}$$

$$L_2(u, v, w, \phi_1, \phi_2) + N_a \frac{\partial^2 v}{\partial x^2} = I_1 v'' + I_2 \phi_2'' \tag{15}$$

$$L_3(u, v, w, \phi_1, \phi_2) + N_a \frac{\partial^2 w}{\partial x^2} = I_1 w'' \tag{16}$$

$$L_4(u, v, w, \phi_1, \phi_2) = I_2 u'' + I_3 \phi_1'' \tag{17}$$

$$L_5(u, v, w, \phi_1, \phi_2) = I_2 v'' + I_3 \phi_2'' \tag{18}$$

By neglecting terms I_2 and I_3 involving in Eqs. (5)-(9) and setting

$$\phi_1 = -\frac{\partial w}{\partial x},\ \phi_2 = -\frac{1}{R}\frac{\partial w}{\partial \theta}, \tag{19}$$

the equations of motion based on a classical shell theory (CST) can be easily obtained.

Here, the two ends of FGM cylindrical shells are considered as simply supported, so that a solution for the motion equations (14)-(18) can be described by

$$\begin{aligned}
u_{mn} &= \overline{A}_{mn} e^{i\omega t} \cos\lambda_m\ x \cos n\theta \\
v_{mn} &= \overline{B}_{mn} e^{i\omega t} \sin\lambda_m\ x \sin n\theta \\
w_{mn} &= \overline{C}_{mn} e^{i\omega t} \sin\lambda_m\ x \cos n\theta \\
\phi_{1mn} &= \overline{H}_{mn} e^{i\omega t} \cos\lambda_m\ x \cos n\theta
\end{aligned}$$

$$\phi_{2mn} = \overline{K}_{mn} e^{i\omega t} \sin \lambda_m \; x \sin n\theta \tag{20}$$

where $\lambda_m = \dfrac{m\pi}{L}$, n represents the number of circumferential waves and m represents he number of axial half-waves.

Substituting Eq. (20) into Eqs.(14)-(18) and letting $N_d = 0$ in Eq.(1), yield

$$\left(\begin{bmatrix} T_{11} & T_{12} & T_{13} & T_{14} & T_{15} \\ T_{21} & T_{12} + \lambda_m^2 N_0 & T_{23} & T_{24} & T_{25} \\ T_{31} & T_{31} & T_{33} + \lambda_m^2 N_0 & T_{34} & T_{35} \\ T_{41} & T_{42} & T_{43} & T_{44} & T_{45} \\ T_{51} & T_{52} & T_{53} & T_{54} & T_{55} \end{bmatrix} - \omega^2 \begin{bmatrix} I_1 & 0 & 0 & I_2 & 0 \\ 0 & I_1 & 0 & 0 & I_2 \\ 0 & 0 & I_1 & 0 & 0 \\ I_2 & 0 & 0 & I_3 & 0 \\ 0 & I_2 & 0 & 0 & I_3 \end{bmatrix} \right) \begin{pmatrix} \overline{A}_{mn} \\ \overline{B}_{mn} \\ \overline{C}_{mn} \\ \overline{H}_{mn} \\ \overline{K}_{mn} \end{pmatrix} = \begin{pmatrix} 0 \\ 0 \\ 0 \\ 0 \\ 0 \end{pmatrix} \tag{21}$$

where T_{ij} is given in appendix A.

The nontrivial solution of the eigen-equation (21) gives the natural frequencies of FGM cylindrical shells under the static axial load, and the corresponding mode shapes. From Eq. (21), the static buckling load N_{0cr} of FGM cylindrical shells can be easily determined by the following equation

$$\begin{vmatrix} T_{11} & T_{12} & T_{13} & T_{14} & T_{15} \\ T_{21} & T_{22} + \lambda_m^2 N_{0cr} & T_{23} & T_{24} & T_{25} \\ T_{31} & T_{32} & T_{33} + \lambda_m^2 N_{0cr} & T_{34} & T_{35} \\ T_{41} & T_{42} & T_{43} & T_{44} & T_{45} \\ T_{51} & T_{52} & T_{53} & T_{54} & T_{55} \end{vmatrix} = 0 \tag{22}$$

The static buckling equation (22) is simplified to

$$a(L,m,n)N_{0cr}^2 + b(L,m,n)N_{0cr} + c(L,m,n) = 0 \quad m,n = 1,2,3,4,\ldots\infty \tag{23}$$

In the above formula, the buckling mode m and n which make N_{0cr} minimum are used to determine the critical buckling load N_{0cr} of the FGM cylindrical shell with given geometry parameters and material parameters.

Neglecting rotary inertias and the transverse shear strains in Eqs.(21), (22) and (23), the corresponding eigen-equation and static buckling equation of FGM cylindrical shells based on a classical shell theory are expressed as

$$\left(\begin{bmatrix} \overline{T}_{11} & \overline{T}_{12} & \overline{T}_{13} \\ T_{21} & \overline{T}_{22}+\lambda_m^2 N_0 & \overline{T}_{23} \\ \overline{T}_{31} & \overline{T}_{32} & \overline{T}_{33}+\lambda_m^2 N_0 \end{bmatrix}\right) - \omega^2 \begin{bmatrix} I_1 & 0 & 0 \\ 0 & I_1 & 0 \\ 0 & 0 & I_1 \end{bmatrix} \begin{Bmatrix} \overline{A}_{mn} \\ \overline{B}_{mn} \\ \overline{C}_{mn} \end{Bmatrix} = \begin{Bmatrix} 0 \\ 0 \\ 0 \end{Bmatrix} \tag{24}$$

$$\begin{vmatrix} \overline{T}_{11} & \overline{T}_{12} & \overline{T}_{13} \\ T_{21} & \overline{T}_{12}+\lambda_m^2 N_0 & \overline{T}_{23} \\ \overline{T}_{31} & \overline{T}_{32} & \overline{T}_{33}+\lambda_m^2 N_0 \end{vmatrix} = 0 \tag{25}$$

$$\overline{a}(L,m,n)N_{0cr}^2 + \overline{b}(L,m,n)N_{0cr} + \overline{c}(L,m,n) = 0 \quad m,n = 1,2,3,4,\ldots\infty \tag{26}$$

$\overline{T}_{ij}$ are given in appendix A.

3. Numerical Results and Discussions

The ceramic material used in this study is silicon nitride and the metal material used is nickel [9]. The density of silicon nitride is ρ_c=2370 kg/m3 and that of nickel is ρ_m=8900 kg/m3, the Poisson's ratio is ν_c=0.24. for silicon nitride and ν_m=0.31 for nickel, which are independent of the temperature. The elastic modules are temperature dependent and are obtained from Ng et al. [9] as

$$E_c = 348.43 \times 10^9 \left(1 - 3.070 \times 10^{-4} T + 2.160 \times 10^{-7} T^2 - 8.946 \times 10^{-11} T^3\right)$$

$$E_m = 223.95 \times 10^9 \left(1 - 2.794 \times 10^{-4} T - 3.998 \times 10^{-9} T^2\right)$$

where E_c and E_m are the elastic modulus of silicon nitride and nickel, respectively, and T is the temperature in Kelvin.The elastic moduli E_c and E_m would be substituted into F_c and F_m in Eq.(3) respectively in order to compute the effective elastic moduli of the FGM shell.

Comparison Studies

Example 1. Based on the classical shell theory, the nondimensionalized natural frequency (λ) for the fully ceramic cylindrical shell with simply supported ends is calculated, which is in a close agreement with Radwan and Genin. [14] shown as Figure 2.

Free Vibration and Buckling Results

Figures 3-5 show the comparison results of nondimensionalized fundamental frequencies $\Omega = 2\omega \times \alpha$; where α the nondimensionalized coefficient is defined as $\alpha = 2\pi R\sqrt{I_1 / A_{11}}$ and ω denotes the fundamental frequency of a silicon nitride-nickel FGM (Type A) cylindrical shell with simply supported ends based on two different shell theories such as the classical shell theory (CST) and the first-order shear deformation theory (FSDT). For cylindrical shells of intermediate length, as are the cases used here and this is conservatively chosen to be in terms of the nickel material, the buckling load is given by Timoshenko and Gere [15]

$$N_{cr} = \frac{E_m h^2}{R\sqrt{3(1 - \nu_m)}} \tag{27}$$

And the static axial load can be nondimensionalized as $\overline{N}_0 = N_0/N_{cr}$ .The effect of transverse shear and rotary inertias on the fundamental frequency can be seen in Figs.3-5. It is observed from Figure3 that the effect of transverse shear, rotary inertias on the fundamental frequency of the FGM (Type A) cylindrical shell is relational to the circumferential mode n. When the circumferential mode n is larger than 3.5, the effect of transverse shear and rotary inertias on the fundamental frequency of the FGM (Type A) cylindrical shell gradually increases as the circumferential mode n increases. Figure4 shows that the difference between the FSDT results considering transverse shear, rotary inertias and the CST results no considering transverse shear, rotary inertias increases as the thickness of the FGM (Type A) cylindrical shell increases. It is concluded that when the ratio h/R of thickness to radius of the FGM (Type A) cylindrical shell is larger, the calculation of the fundamental frequency of the shell under extensional loading should be based on the FSDT method considering transverse shear, rotary inertias.

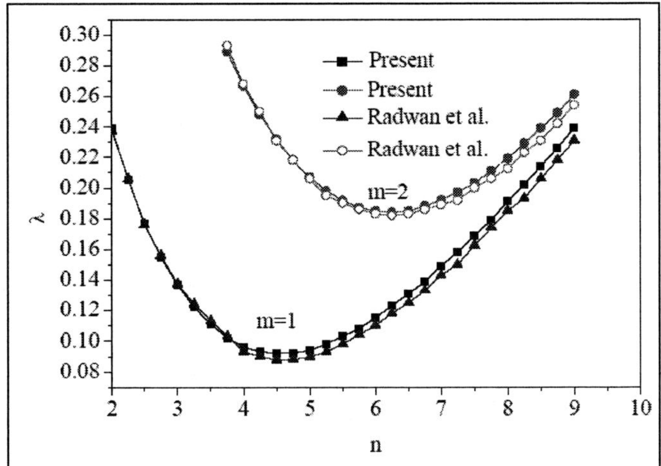

Figure 2. Non-dimensional fundamental frequency λ versus circumferential modes (m,n) for simply supported cylindrical shell.

$$\left[h/R = 0.01, L/R = 2.5, \lambda^2 = \left(I_1 R^2/E\right)\omega^2, E = E_c h/(1-\nu_c^2), N_0 = 0\right].$$

ω denotes the fundamental frequency.

The fundamental frequency increases with an increase in the thickness to radius ratio h/R under extensional loading. This is due to the fact that the cylindrical shell has larger stiffness as the ratio h/R of thickness to radius increases.

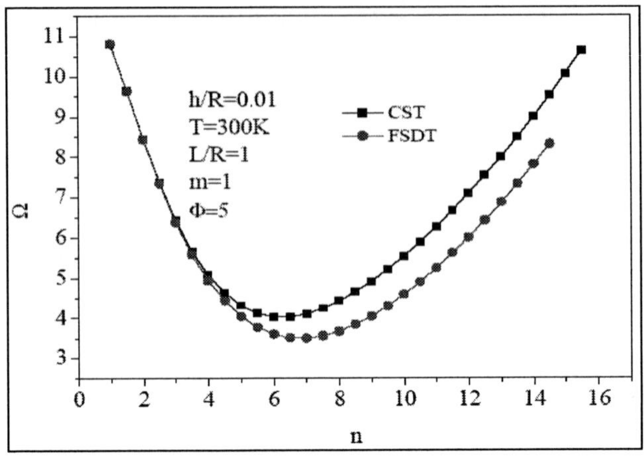

Figure 3. Non-dimensional fundamental frequency Ω versus circumferential mode n for a simply supported silicon nitride-nickel FGM Type A cylindrical shell under axial extensional loading, based on two different solving methods.

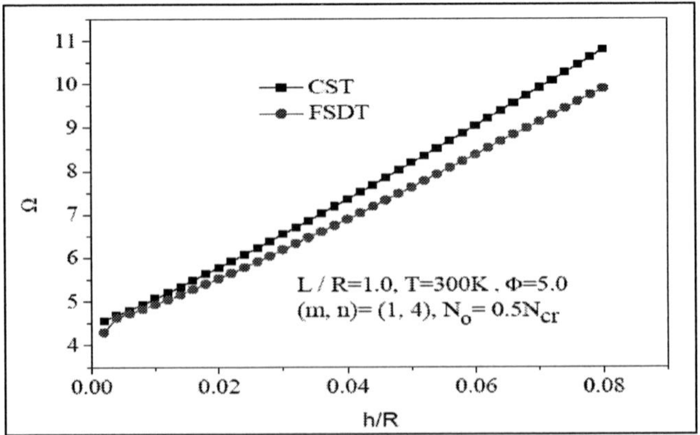

Figure 4. Non-dimensional fundamental frequency Ω versus thickness to radius ratio h/R for a simply supported silicon nitride-nickel FGM Type A cylindrical shell under axial extensional loading, based on two different solving methods.

Figure 5 shows that the fundamental frequency of the FGM (Type A) cylindrical shell decreases as the static axial compressive load increases. This is consistent with the results reported by Lam and Ng [16] for the dynamic stability analysis of isotropic cylindrical shells. It is also shown that relatively higher values of fundamental frequencies will be resulted in employing CST compared with that of FSDT. This reveals the influence of transverse shear and rotary inertias.

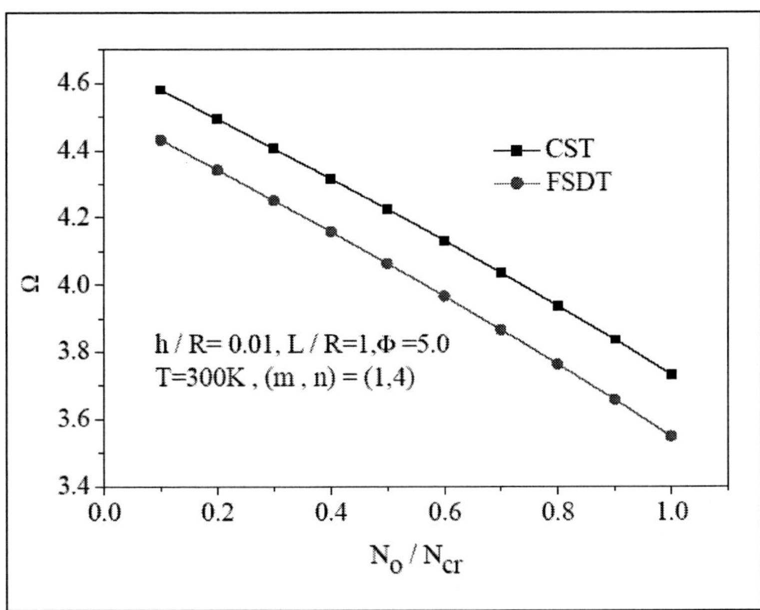

Figure 5. Non-dimensional fundamental frequency Ω versus non-dimensional axial compressive load N_o/N_{cr} for a simply supported silicon nitride-nickel FGM (Type A) cylindrical shell, based on two different solving methods.

Figures 6-8 shows the numerical results of a silicon nitride-nickel FGM Type A cylindrical shell with simply supported ends, based on to the first-order shear deformation theory (FSDT). From Figure6, it is shown that the effect of volume fraction exponent on nondimensionalized natural frequency is obvious only when the volume fraction exponent Φ is less than 5. It is observed that when the volume fraction exponent Φ is larger 8, the results are already very close to those associated with $\Phi = \infty$.

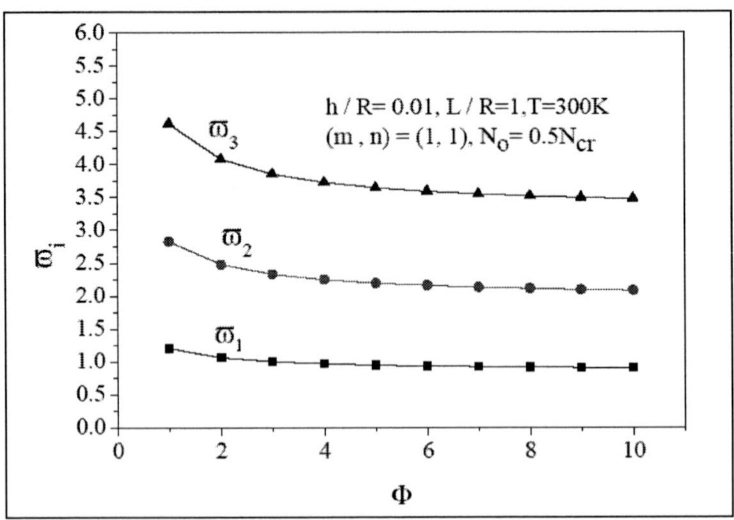

Figure 6. Non-dimensional natural frequency $\varpi_i = \omega_i(\Phi)/\omega_1(10)\ (i=1,2,3)$ versus volume fraction exponent for a simply supported silicon nitride-nickel FGM Type A cylindrical shell under axial extensional loading [$\omega_1(10)$ denotes the fundamental frequency for Φ=1and $\omega_i(\Phi)\ (i=1,2,3)$ denotes the three lowest natural frequency].

For Type A material, when $\Phi = 0$, the shell is fully ceramic (silicon nitrate); when $\Phi = \infty$, the shell is fully metal (nickel); the nondimensionalized natural frequency decrease slowly as the volume fraction exponent increases. Figure7 shows the effect of thermal environments on nondimensionalized natural frequency. It is seen from Figure7 that the natural frequency decrease as temperature rises, and the effect of thermal environments on the third order frequency is larger than the first order frequency

Figure 8(a) shows the effect of the ratio of length to radius L/R on nondimensionalized buckling load N_{ocr}/N_{cr} . Figure 8(b) shows that the buckling mode (m,n) is dependent on the ratio of length to radius L/R of the silicon nitride-nickel FGM cylindrical shell. It is clear from the results that when the length of the cylindrical shells increases, buckling load decreases and the numbers of the buckling mode are small.

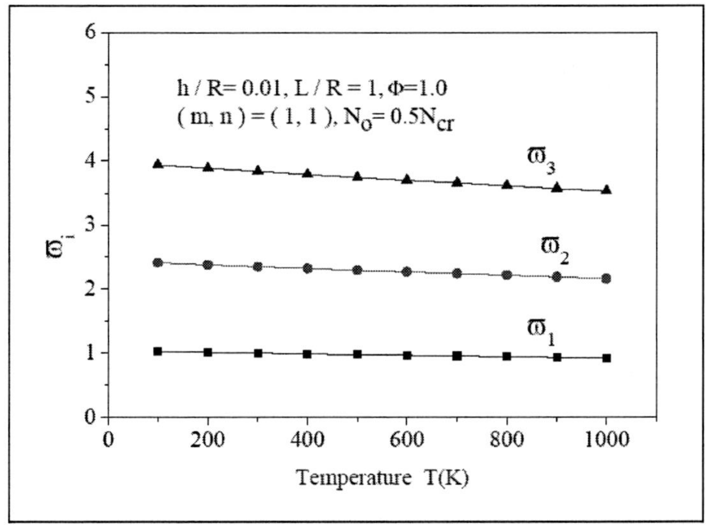

a)

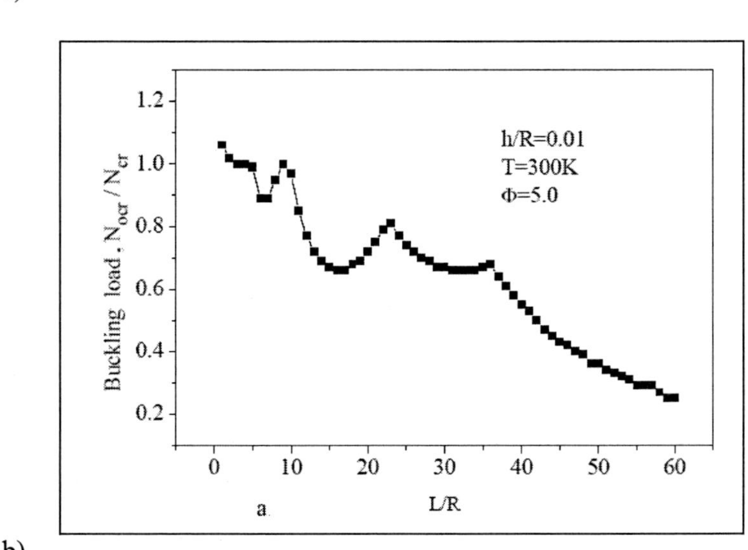

b)

Figure 7. Effect of thermal environments on the non-dimensional natural frequency, $\varpi_i = \omega_i(T)/\omega_1(300)$ for a simply supported silicon nitride-nickel FGM Type A cylindrical shell under axial extensional loading [$\omega_1(300)$ denotes the fundamental frequency for T=300K and $\omega_i(T)$ $(i = 1,2,3)$ denotes the three lowest natural frequency].

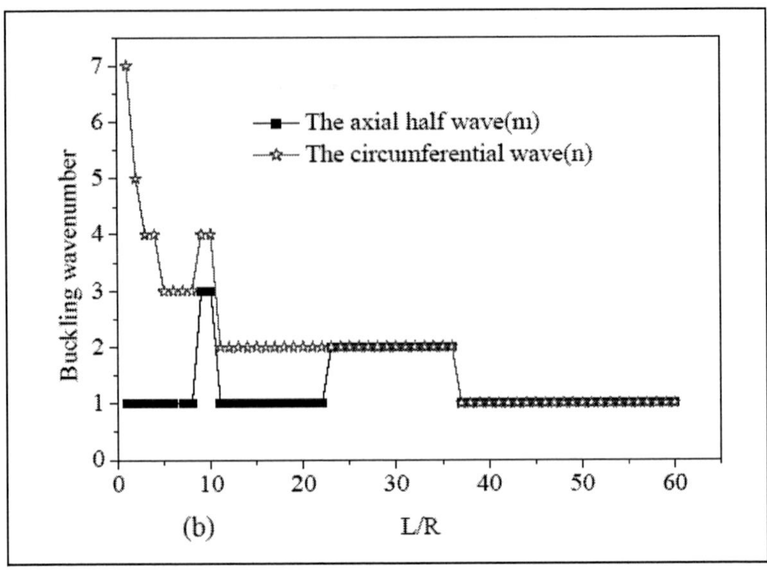

Figure 8. (a) Non-dimensional buckling load N_{ocr}/N_{cr} versus length to radius ratio L/R for a simply supported silicon nitride-nickel FGM (Type A) cylindrical shell; (b) Buckling mode (m,n) corresponding to the length to radius ratio L/R for a simply supported silicon nitride-nickel FGM (Type A) cylindrical shell.

CONCLUSIONS

This chapter reports the result of an investigation into the effect of the first-order shear deformation theory considering rotary inertia and the transverse shear strains on the free vibration and buckling of functionally graded cylindrical shells with simply-supported ends under combined static and periodic axial forces. The result obtained using the first-order shear deformation theory (FSDT) is compared with that obtained using classical shell theory (CST). The differences between the results from FSDT and the results from CST increase as the deformation mode and thickness increase. It was found that reasonable control can be achieved on the natural frequencies and buckling load by correctly varying the ratio of length to radius, the ratio of thickness to radius, the amplitude of static axial load, thermal environment and volume fraction exponent.

APPENDIX A

$$T_{11} = A_{11}\lambda_m^2 + \frac{A_{66}n^2}{R^2} \qquad T_{12} = -\frac{A_{12} + A_{66}}{R}\lambda_m n \qquad T_{13} = -\frac{A_{12}}{R}\lambda_m$$

$$T_{14} = B_{66}\lambda_m^2 + \frac{B_{66}n^2}{R^2} \qquad T_{15} = -\frac{B_{12} + B_{66}}{R}\lambda_m n \qquad T_{21} = -\frac{A_{12} + A_{66}}{R}\lambda_m n$$

$$T_{22} = A_{66}\lambda_m^2 + \frac{A_{22}n^2}{R^2} + \frac{C_{55}}{R^2} \qquad T_{23} = \frac{A_{22} + C_{55}}{R}n \qquad T_{24} = -\frac{B_{21} + B_{66}}{R}\lambda_m n$$

$$T_{25} = B_{66}\lambda_m^2 + \frac{B_{22}n^2}{R^2} - \frac{C_{55}}{R^2} \qquad T_{31} = -\frac{A_{21}}{R}\lambda_m \qquad T_{32} = \frac{C_{55} + A_{22}}{R^2}n$$

$$T_{33} = C_{44}\lambda_m^2 + \frac{C_{55}n^2}{R^2} + \frac{A_{22}}{R^2} \qquad T_{34} = \left(C_{44} - \frac{B_{21}}{R}\right)\lambda_m \qquad T_{35} = -\frac{C_{55}R - B_{22}}{R^2}n$$

$$T_{41} = B_{11}\lambda_m^2 + \frac{B_{66}n^2}{R^2} \qquad T_{42} = -\frac{B_{12} + B_{66}}{R}\lambda_m n \qquad T_{43} = \left(C_{44} - \frac{B_{12}}{R}\right)\lambda_m$$

$$T_{44} = D_{11}\lambda_m^2 + \frac{D_{66}n^2}{R^2} + C_{44} \qquad T_{45} = -\frac{D_{12} + D_{66}}{R}\lambda_m n \qquad T_{51} = -\frac{B_{21} + B_{66}}{R}\lambda_m n$$

$$T_{52} = B_{66}\lambda_m^2 + \frac{B_{22}n^2}{R^2} - \frac{C_{55}}{R} \qquad T_{53} = \frac{B_{22} - RC_{55}}{R^2}n \qquad T_{55} = \frac{D_{22}}{R^2}n^2 + C_{55} + \frac{D_{66}}{R}\lambda_m^2$$

$$T_{54} = -\frac{D_{21}}{R}\lambda_m n - \frac{D_{66}}{R}\lambda_m n \qquad \overline{T}_{11} = A_{11}\lambda_m^2 + \frac{A_{66}n^2}{R^2} \qquad \overline{T}_{21} = \left[-\frac{A_{12}+A_{66}}{R} - \frac{B_{66}+B_{12}}{R^2}\right]\lambda_m n$$

$$\overline{T}_{22} = A_{66}\lambda_m^2 + \frac{A_{22}n^2}{R^2} + \frac{B_{66}}{R}\lambda_m^2 + \frac{B_{22}n^2}{R^3} \qquad \overline{T}_{13} = -\frac{A_{12}}{R}\lambda_m - B_{11}\lambda_m^3 - \frac{B_{12}}{R^2}\lambda_m n^2 - 2\frac{B_{66}}{R^2}\lambda_m n^2$$

$$\overline{T}_{23} = \left[\left(\frac{2B_{66}}{R} + \frac{B_{12}}{R} + \frac{2D_{66}}{R^2} + \frac{D_{12}}{R}\right)\lambda_m^2 n + \frac{A_{22}}{R^2}n - \frac{B_{22}}{R^3}n^3 + \frac{B_{22}}{R^3}n - \frac{D_{22}}{R^4}n^3\right]$$

$$\overline{T}_{31} = \left[-B_{11}\lambda_m^3 - \left(\frac{2B_{66}+B_{12}}{R^2}\right)\lambda_m n^2 - \frac{A_{12}}{R}\lambda_m\right] \qquad \overline{T}_{12} = -\frac{A_{12}+A_{66}}{R}\lambda_m n$$

$$\overline{T}_{32} = \left[(2B_{66}+B_{12})\frac{\lambda_m^2 n}{R} + \frac{B_{22}}{R^3}n^3 + \frac{A_{22}}{R^2}n\right]$$

$$\overline{T}_{33} = \left[\frac{2D_{12}+4D_{66}}{R^2}\lambda_m^2 n^2 + \frac{2B_{12}}{R}\lambda_m^2 + \frac{2B_{22}}{R^3}n^2 + \frac{A_{22}}{R^2} + D_{11}\lambda_m^4 + \frac{D_{22}}{R^4}n^4\right]$$

REFERENCES

[1] Koizumi, M. and Niino, M., Overview of FGM research in Japan, *MRS Bull.*;1995; 20:, 19– 21.

[2] Reddy J.N. Mechanics of Laminated Composite Plates and Shells, Second Edition ,CRC Press, New York;2004.

[3] Reddy, J.N. and Chin, C. D., Thermomechanical Analysis of Functionally Graded Cylinders and Plates. *J. Thermal Stresses* 1998; 26(1): 593-626.

[4] Arciniega R. A., and Reddy, J.N., Large deformation analysis of functionally graded shells. *International Journal of Solids and Structures* 2007; 44: 2036-2052.

[5] Praveen, G. N., Chin, C. D. and Reddy, J.N., Thermoelastic Analysis of a Functionally Graded Ceramic-Metal Cylinder. *ASCE Journal of Engineering Mechanics* 1999;125(11): 1259-1267.

[6] Loy CT, Lam KY and Reddy JN. Vibration of functionally graded cylindrical shells, *International Journal of Mechanical Sciences* 1999;41: 309-24.

[7] Pradhan SC, Loy CT, Lam KY. and Reddy JN. Vibration characteristics of functionally graded cylindrical shells under various boundary conditions, *Applied Acoustics* 2000; 61; 119-29.

[8] Gong SW, Lam KY and Reddy JN. The elastic response of functionally graded cylindrical shells to low-velocity impact. *International Journal of Impact Engineering* 1999; 22: 397-417.

[9] Ng TY, Lam KM, Liew KM and Reddy JN. Dynamic stability analysis of functionally graded cylindrical shells under periodic axial loading. *International Journal of Solids and Structures* 2001; 38: 1295-309.

[10] Jafari AA, Khalili SMR and Azarafza R. Transient dynamic response of composite circular cylindrical shells under radial impulse load and axial compressive loads. *Thin-Walled Structures* 2005; 43:1763-86.

[11] Bhangale RK and Ganesan N. Free vibration studies of simply supported non-homogeneous functionally graded magneto-electro-elastic finite cylindrical shells. *Journal of Sound and Vibration* 2005; 288: 412–22.

[12] Ravikiran K and Ganesan N. Buckling and free vibration analysis of functionally graded cylindrical shells subjected to a temperature-specified boundary condition. *Journal of Sound and Vibration* 2006; 289: 450-80.

[13] Timoshenko SP and Woinowsky-Krieger S. Theory of Plates and Shells. Second Edition, McGraw-Hill, New York; 1959.

[14] Radwan HR and Genin J. Dynamic instability in cylindrical shells. *Journal of Sound and Vibration* 1978; 56:373-82.

[15] Timoshenko SP and Gere JM . Theory of Elastic Stability. McGraw-Hill, New York; 1961.

[16] Lam KY and Ng TY. Dynamic stability of cylindrical shells subjected to conservative periodic axial loads using different shell theories. *Journal of Sound and Vibration* 1997; 207:497-520.

Chapter 2

STABILITY ANALYSIS OF FGM CYLINDRICAL SHELLS UNDER COMBINED STATIC AND PERIODIC AXIAL FORCES

ABSTRACT

The effect of transverse shear and rotary inertias on the dynamic stability of functionally graded cylindrical shells subjected to combined static and periodic axial forces is investigated in this chapter. Material properties of functionally graded cylindrical shells are considered temperature-dependent and are graded in the thickness direction according to a power-law distribution in terms of the volume fractions of the constituents. Numerical results for silicon nitride-nickel cylindrical shells are presented based on two different methods: the first-order shear deformation theory (FSDT) which considers the transverse shear strains and the rotary inertias, and the classical shell theory (CST). The results obtained show that the effect of transverse shear and rotary inertias on the dynamic stability of functionally graded cylindrical shells subjected to combined static and periodic axial forces is dependent on the shell's material composition, environmental temperature, amplitude of static load, deformation mode, and the shell's geometry parameters.

1. INTRODUCTION

As stated in previous chapter functionally graded materials (FGMs) are microscopically inhomogeneous materials. By choosing specific manufacturing processes, the properties of the produced FGMs may vary from point to point or from layer to layer. Most importantly, certain properties (usually the desired ones) of the specifically produced FGMs are superior to those of corresponding homogeneous materials. Recent advances in manufacturing technology have made FGMs more preferential from both functional and economic points of view. As a result, the usage of FGMs has been significantly broadened. For example, there are reports of early successful applications of FGMs used as high-temperature materials in nuclear reactors and chemical plants in Japan [1]. Now, FGMs are also being considered as potential structural materials for future high-speed spacecrafts. Formulation and theoretical analysis of the FGM plates and shells were presented by Reddy [2

Despite the evident importance in practical applications, investigations on the static and dynamic characteristics of FGM shell structures are still limited in number. Among those available, Using Love's shell theory and the Galerkin method, Sofiyev [3] presented an analytic solution for the stability behavior of cylindrical shells made of compositionally (or functionally) graded ceramic–metal materials under axial compressive loads. Also, a solution of the dynamic stability of functionally graded shells under periodic axial loading using the Galerkin procedure was presented by Darabi et al [4], while Matsunaga presented a free vibration and stability analysis of functionally graded circular and shallow shells according to a 2D higher order deformation theory [5,6]. This chapter studies the effect of transverse shear and rotary inertias on the dynamic stability of functionally graded cylindrical shells subjected to combined static and periodic axial forces through a comparison of results obtained by using the following methods: the first-order shear deformation theory (FSDT) which considers transverse shear strains and rotary inertias, and the classical shell theory (CST). Material properties of functionally graded cylindrical shells are considered temperature-dependent and graded in the thickness direction according to a power-law distribution in terms of the volume fractions of the constituents. The results show that the effect of transverse shear and rotary inertias on the dynamic stability of functionally graded cylindrical shells subjected to combined static and periodic axial forces is dependent on the shell's material composition, environmental temperature, amplitude of static load, deformation mode, and the shell's geometry parameters. It is found that the effect of transverse shear and rotary inertias on the dynamic stability of functionally graded cylindrical shells subjected to combined static and periodic axial forces cannot be neglected. The new features of the effect of transverse shear and rotary inertias on the dynamic stability of functionally graded cylindrical shells.

2. Theoretical Formulations

Like previous chapter an FGM cylindrical shell with mean radius of R, thickness h, and length L is shown in Figure 1 of chapter 1. The displacement components in the x, θ, and z directions are denoted by u, v, and w, respectively. The pulsating axial load is given by

$$N_a = N_o + N_d \cos Pt \,, \tag{1}$$

where P is the frequency of excitation in radians per unit time.

The temperature and position-dependent material properties of FGM cylindrical shells are accurately modeled by using a simple rule of mixtures for the stiffness parameters coupled with the temperature-dependent properties of the constituents. The volume fraction is described by a spatial function as follows:

$$V(z) = (z/h + 1/2)^{\Phi} \quad , \quad (0 \leq \Phi \leq \infty) \,, \tag{2}$$

where Φ expresses the volume fraction exponent. The combination of these functions gives rise to the effective properties of functionally graded materials. An FGM cylindrical shell that is metal-rich at the inner surface and ceramic-rich at the outer surface is defined as Type A. The corresponding effective material properties are expressed as

$$F_{eff}(T,z) = F_c(T)V(z) + F_m(T)(1 - V(z)) \,, \tag{3}$$

where F_{eff} is the effective material property of the FGM cylindrical shell, including the effective elastic modulus, effective mass density, and effective Poisson's ratio. F_c and F_m are the temperature-dependent properties of the ceramic and metal, respectively . On the other hand, an FGM cylindrical shell that is ceramic-rich at the inner surface and metal-rich at the outer surface is defined as Type B, whose effective material properties are given by

$$F_{eff}(T,z) = F_m(T)V(z) + F_c(T)(1 - V(z)) \,. \tag{4}$$

Based on the first-order shear deformation theory (FSDT), the equations of motion for an FGM cylindrical shell under axially dynamic load are as follows [3, 7]:

$$\frac{\partial N_1}{\partial x}+\frac{1}{R}\frac{\partial N_6}{\partial \theta}=I_1 u''+I_2 \phi_1'' \tag{5}$$

$$\frac{\partial N_6}{\partial x}+\frac{1}{R}\frac{\partial N_1}{\partial \theta}+\frac{1}{R}Q_2+N_a\frac{\partial^2 v}{\partial x^2}=I_1 v''+I_2 \phi_2'' \tag{6}$$

$$\frac{\partial Q_1}{\partial x}+\frac{1}{R}\frac{\partial Q_2}{\partial \theta}-\frac{N_2}{R}+N_a\frac{\partial^2 w}{\partial x^2}=I_1 w'' \tag{7}$$

$$\frac{\partial M_1}{\partial x}+\frac{1}{R}\frac{\partial M_6}{\partial \theta}-Q_1=I_2 u''+I_3 \phi_1'' \tag{8}$$

$$\frac{\partial M_6}{\partial x}+\frac{1}{R}\frac{\partial M_2}{\partial \theta}-Q_2=I_2 v''+I_3 \phi_2'' \tag{9}$$

where ϕ_1 and ϕ_2 are the rotations of a normal to the reference surface, $I_i (i=1,2,3)$ is the mass inertia terms defined as

$$\left(I_1, I_2, I_3\right)=\int_{-\frac{h}{2}}^{\frac{h}{2}}\rho(z)(1,z,z^2)dz\ , \tag{10}$$

and $\rho(z)$ is the effective mass density of functionally graded materials.

The stress resultants of FGM cylindrical shells are given by .

$$\begin{pmatrix} N_1 \\ N_2 \\ N_6 \\ M_1 \\ M_2 \\ M_6 \end{pmatrix} = \begin{pmatrix} A_{11} & A_{12} & 0 & B_{11} & B_{12} & 0 \\ A_{21} & A_{22} & 0 & B_{21} & B_{22} & 0 \\ 0 & 0 & A_{66} & 0 & 0 & B_{66} \\ B_{11} & B_{12} & 0 & D_{11} & D_{12} & 0 \\ B_{21} & B_{22} & 0 & D_{21} & D_{22} & 0 \\ 0 & 0 & B_{66} & 0 & 0 & D_{66} \end{pmatrix} \begin{pmatrix} \varepsilon_1 \\ \varepsilon_2 \\ \varepsilon_6 \\ \kappa_1 \\ \kappa_2 \\ \kappa_6 \end{pmatrix}$$

$$\begin{pmatrix} Q_1 \\ Q_2 \end{pmatrix} = \begin{pmatrix} C_{44} & 0 \\ 0 & C_{55} \end{pmatrix} \begin{pmatrix} \varepsilon_5 \\ \varepsilon_4 \end{pmatrix} \tag{11}$$

where A_{ij}, B_{ij}, D_{ij} and C_{ij} are, respectively, the extensional, coupling, bending, and shear stiffness. These are given by

$$\left(A_{ij}, B_{ij}, D_{ij}\right) = \int_{-\frac{h}{2}}^{\frac{h}{2}} Q_{ij}(1, z, z^2)dz\,, \quad (i, j = 1,2,6)$$

$$Q_{11} = Q_{22} = \frac{E_{eff}}{1 - \nu_{eff}{}^2}$$

$$Q_{12} = Q_{21} = \frac{\nu_{eff} E_{eff}}{A(1 - \nu_{eff}{}^2)}$$

$$Q_{44} = \kappa \frac{E_{eff}}{2\left(1 + \nu_{eff}\right)}$$

$$Q_{55} = Q_{44} / A$$

$$C_{44} = \int_{-\frac{h}{2}}^{\frac{h}{2}} Q_{44} dz$$

$$C_{55} = \int_{-\frac{h}{2}}^{\frac{h}{2}} Q_{55} dz$$

$$Q_{66} = \frac{E_{eff}}{2A(1+\nu_{eff})}$$
$$A = 1 + z/R \tag{12}$$

where E_{eff} and ν_{eff} are the effective elastic modulus and effective Poisson's ratio of FGM cylindrical shells, respectively. κ is the shear correction factor introduced by Reddy [8] and is equal to 5/6. The strains are expressed as

$$\varepsilon_1 = \frac{\partial u}{\partial x}, \quad \varepsilon_2 = \frac{1}{R}\left(\frac{\partial v}{\partial \theta} + w\right), \quad \varepsilon_6 = \frac{\partial v}{\partial x} + \frac{1}{R}\frac{\partial u}{\partial \theta}, \quad \varepsilon_4 = \phi_2 + \frac{1}{R}\frac{\partial w}{\partial \theta},$$
$$\varepsilon_5 = \phi_1 + \frac{\partial w}{\partial x}, \quad \kappa_1 = \frac{\partial \phi_1}{\partial x}, \quad \kappa_2 = \frac{1}{R}\frac{\partial \phi_2}{\partial \theta}, \quad \kappa_6 = \frac{\partial \phi_2}{\partial x} + \frac{1}{R}\frac{\partial \phi_1}{\partial \theta} \tag{13}$$

Utilizing Eqns.(5-9), (11) and (13), the equations of motion can be expressed in terms of generalized displacement $(u, v, w, \phi_1, \phi_2)$ as follows:

$$L_1(u, v, w, \phi_1, \phi_2) = I_1 u'' + I_2 \phi_1'' \tag{14}$$

$$L_2(u, v, w, \phi_1, \phi_2) + N_a \frac{\partial^2 v}{\partial x^2} = I_1 v'' + I_2 \phi_2'' \tag{15}$$

$$L_3(u, v, w, \phi_1, \phi_2) + N_a \frac{\partial^2 w}{\partial x^2} = I_1 w'' \tag{16}$$

$$L_4(u, v, w, \phi_1, \phi_2) = I_2 u'' + I_3 \phi_1'' \tag{17}$$

$$L_5(u, v, w, \phi_1, \phi_2) = I_2 v'' + I_3 \phi_2'' \tag{18}$$

By neglecting terms I_2 and I_3 involved in Eqns. (5-9) and setting

$$\phi_1 = -\frac{\partial w}{\partial x}, \quad \phi_2 = -\frac{1}{R}\frac{\partial w}{\partial \theta}, \tag{19}$$

the equations of motion based on a classical shell theory (CST) can be easily obtained.

Here, the two ends of the FGM cylindrical shells are considered simply supported, so a solution for the motion equations (14-18) can be described by

$$\begin{aligned}
u_{mn} &= \overline{A}_{mn} e^{i\omega t} \cos\lambda_m\ x \cos n\theta \\
v_{mn} &= \overline{B}_{mn} e^{i\omega t} \sin\lambda_m\ x \sin n\theta \\
w_{mn} &= \overline{C}_{mn} e^{i\omega t} \sin\lambda_m\ x \cos n\theta \\
\phi_{1mn} &= \overline{H}_{mn} e^{i\omega t} \cos\lambda_m\ x \cos n\theta \\
\phi_{2mn} &= \overline{K}_{mn} e^{i\omega t} \sin\lambda_m\ x \sin n\theta
\end{aligned} \tag{20}$$

where $\lambda_m = \dfrac{m\pi}{L}$, n represents the number of circumferential waves, and m represents the number of axial half-waves.

Substituting Eqn. (20) into Eqns. (14-18) and letting $N_d = 0$ in Eq.(1) yields

$$\left(\begin{bmatrix} T_{11} & T_{12} & T_{13} & T_{14} & T_{15} \\ T_{21} & T_{12}+\lambda_m^2 N_0 & T_{23} & T_{24} & T_{25} \\ T_{31} & T_{31} & T_{33}+\lambda_m^2 N_0 & T_{34} & T_{35} \\ T_{41} & T_{42} & T_{43} & T_{44} & T_{45} \\ T_{51} & T_{52} & T_{53} & T_{54} & T_{55} \end{bmatrix} - \omega^2 \begin{bmatrix} I_1 & 0 & 0 & I_2 & 0 \\ 0 & I_1 & 0 & 0 & I_2 \\ 0 & 0 & I_1 & 0 & 0 \\ I_2 & 0 & 0 & I_3 & 0 \\ 0 & I_2 & 0 & 0 & I_3 \end{bmatrix}\right) \begin{pmatrix} \overline{A}_{mn} \\ \overline{B}_{mn} \\ \overline{C}_{mn} \\ \overline{H}_{mn} \\ \overline{K}_{mn} \end{pmatrix} = \begin{pmatrix} 0 \\ 0 \\ 0 \\ 0 \\ 0 \end{pmatrix} \tag{21}$$

where T_{ij} is given in Appendix A of chapter 1.

To solve the equations of motion containing the dynamic load N_d, a solution is sought in the form shown below:

$$u=\sum_{j=1}^{5}\sum_{m=1}^{\infty}\sum_{n=1}^{\infty}q_{mnj}(t)U_{mnj}(x,\theta)=\sum_{j=1}^{5}\sum_{m=1}^{\infty}\sum_{n=1}^{\infty}q_{mnj}(t)\overline{A}_{mnj}\cos\lambda_m\, x\cos n\theta \tag{22}$$

$$v=\sum_{j=1}^{5}\sum_{m=1}^{\infty}\sum_{n=1}^{\infty}q_{mnj}(t)V_{mnj}(x,\theta)=\sum_{j=1}^{5}\sum_{m=1}^{\infty}\sum_{n=1}^{\infty}q_{mnj}(t)\overline{B}_{mnj}\sin\lambda_m\, x\sin n\theta \tag{23}$$

$$w=\sum_{j=1}^{5}\sum_{m=1}^{\infty}\sum_{n=1}^{\infty}q_{mnj}(t)W_{mnj}(x,\theta)=\sum_{j=1}^{5}\sum_{m=1}^{\infty}\sum_{n=1}^{\infty}q_{mnj}(t)\overline{C}_{mnj}\sin\lambda_m\, x\cos n\theta \tag{24}$$

$$\phi_1=\sum_{j=1}^{5}\sum_{m=1}^{\infty}\sum_{n=1}^{\infty}q_{mnj}(t)\theta_{xmnj}(x,\theta)=\sum_{j=1}^{5}\sum_{m=1}^{\infty}\sum_{n=1}^{\infty}q_{mnj}(t)\overline{H}_{mnj}\cos\lambda_m\, x\cos n\theta \tag{25}$$

$$\phi_2=\sum_{j=1}^{5}\sum_{m=1}^{\infty}\sum_{n=1}^{\infty}q_{mnj}(t)\theta_{\theta mnj}(x,\theta)=\sum_{j=1}^{5}\sum_{m=1}^{\infty}\sum_{n=1}^{\infty}q_{mnj}(t)\overline{K}_{mnj}\sin\lambda_m\, x\sin n\theta \tag{26}$$

where $q_{mnj}(t)$ is a generalized co-ordinate and U_{mnj}, V_{mnj}, W_{mnj}, θ_{xmnj} ,and $\theta_{\theta mnj}$ are the modal functions of FGM cylindrical shells with simply supported ends under the axially static load N_0. Substituting Eqns. (22-26) into Eqns. (14-18), yields

$$L_1(U_{mnj},V_{mnj},W_{mnj},\theta_{xmnj},\theta_{\theta mnj})=-I_1\omega_{mnj}^2U_{mnj}-I_2\omega_{mnj}^2H_{mnj} \tag{27}$$

$$L_2(U_{mnj},V_{mnj},W_{mnj},\theta_{xmnj},\theta_{\theta mnj})-N_0\lambda_m^2V_{mnj}=-I_1\omega_{mnj}^2V_{mnj}-I_2\omega_{mnj}^2\theta_{\theta mnj} \tag{28}$$

$$L_3(U_{mnj}, V_{mnj}, W_{mnj}, \theta_{xmnj}, \theta_{\theta mnj}) - N_0\lambda_m^2 W_{mnj} = -I_1\omega_{mnj}^2 W_{mnj} \tag{29}$$

$$L_4(U_{mnj}, V_{mnj}, W_{mnj}, \theta_{xmnj}, \theta_{\theta mnj}) = -I_2\omega_{mnj}^2 U_{mnj} - I_3\theta_{xmnj} \tag{30}$$

$$L_5(U_{mnj}, V_{mnj}, W_{mnj}, \theta_{xmnj}, \theta_{\theta mnj}) = -I_2\omega_{mnj}^2 V_{mnj} - I_3\theta_{\theta mnj} \tag{31}$$

Eqns. (14-18) may be rewritten as

$$\sum_{j=1}^{5}\sum_{m=1}^{\infty}\sum_{n=1}^{\infty}\left(q''_{mnj} + \omega_{mnj}^2 q_{mnj}\right)\left(I_1 A_{mnj} + I_2 H_{mnj}\right)\cos\lambda_m\ x\cos n\theta = 0 \tag{32}$$

$$\sum_{j=1}^{5}\sum_{m=1}^{\infty}\sum_{n=1}^{\infty}\left(q''_{mnj} + \omega_{mnj}^2 q_{mnj}\right)\left(I_1 B_{mnj} + I_2 K_{mnj}\right)\sin\lambda_m\ x\sin n\theta$$
$$+ N_d\cos Pt\sum_{j=1}^{5}\sum_{m=1}^{\infty}\sum_{n=1}^{\infty}\lambda_m^2 B_{mnj} q_{mnj}\sin\lambda_m\ x\sin n\theta = 0 \tag{33}$$

$$\sum_{j=1}^{5}\sum_{m=1}^{\infty}\sum_{n=1}^{\infty}\left(q''_{mnj} + \omega_{mnj}^2 q_{mnj}\right) I_1 C_{mnj}\sin\lambda_m\ x\cos n\theta$$
$$+ N_d\cos Pt\sum_{j=1}^{5}\sum_{m=1}^{\infty}\sum_{n=1}^{\infty}\lambda_m^2 C_{mnj} q_{mnj}\sin\lambda_m\ x\cos n\theta = 0 \tag{34}$$

$$\sum_{j=1}^{5}\sum_{m=1}^{\infty}\sum_{n=1}^{\infty}\left(q''_{mnj} + \omega_{mnj}^2 q_{mnj}\right)\left(I_2 A_{mnj} + I_3 H_{mnj}\right)\cos\lambda_m\ x\cos n\theta = 0 \tag{35}$$

$$\sum_{j=1}^{5}\sum_{m=1}^{\infty}\sum_{n=1}^{\infty}\left(q''_{mnj}+\omega_{mnj}^{2}q_{mnj}\right)\left(I_2B_{mnj}+I_3K_{mnj}\right)\sin\lambda_m\ x\sin n\theta=0 \tag{36}$$

Making use of the orthogonality condition, Eqns .(32-36) are simplified to

$$\begin{bmatrix} m_1 & 0 & 0 & 0 \\ 0 & m_1 & 0 & 0 \\ 0 & 0 & \mathrm{O} & 0 \\ 0 & 0 & 0 & m_{5N^2} \end{bmatrix}\begin{Bmatrix} q''_1 \\ q''_2 \\ M \\ q''_{5N^2} \end{Bmatrix}+$$

$$\left(\begin{bmatrix} k_1 & 0 & 0 & 0 \\ 0 & k_2 & 0 & 0 \\ 0 & 0 & \mathrm{O} & 0 \\ 0 & 0 & 0 & k_{5N^2} \end{bmatrix}-N_d\cos Pt\begin{bmatrix} \overline{Q}_1 & 0 & 0 & 0 \\ 0 & \overline{Q}_2 & 0 & 0 \\ 0 & 0 & \mathrm{O} & 0 \\ 0 & 0 & 0 & \overline{Q}_{5N^2} \end{bmatrix}\right)\begin{Bmatrix} q_1 \\ q_2 \\ M \\ q_{5N^2} \end{Bmatrix}=0 \tag{37}$$

where

$$m_I=\frac{\pi l}{2}\left[\left(I_1\overline{B}_I+I_2\overline{K}_I\right)^2+\left(I_1\overline{A}_I+I_2\overline{H}_I\right)^2+\left(I_1\overline{C}_I\right)^2+\left(I_2\overline{A}_I+I_3\overline{H}_I\right)^2+\left(I_2\overline{B}_I+I_3\overline{K}_I\right)^2\right]$$

$$k_I=\omega_I^2m_I,\quad \overline{Q}_I=-\frac{\pi L}{2}\lambda_m^2\left[\overline{B}_I\left(I_1\overline{B}_I+I_2\overline{K}_I\right)+I_1\overline{C}_I{}^2\right] \tag{38}$$

$(m,n)=(1,1):\quad I=1(j=1),\ 2(j=2),\quad 3(j=3),\quad 4(j=4),\quad 5(j=5);$

$(m,n)=(1,2):\quad I=6(j=1),\ 7(j=2),\quad 8(j=3),\quad 9(j=4),\quad 10(j=5);$

$(m,n)=(1,N):\quad I=5(N-4)\ (j=1),\ 5(N-3)\ (j=2),\quad 5N\ (j=5);$

$(m,n)=(2,1):\quad I=5N+1\ (j=1),\ 5N+2\ (j=2),\quad 5N+5\ (j=5);$

$(m,n)=(N,N):\quad I=5N^2-4\ (j=1),\ 5N^2-3\ (j=2),\quad 5N^2\ (j=5);$

(39)

In the above formula, the coefficients of mode shapes $\overline{A}_I$, $\overline{B}_I$, $\overline{C}_I$, $\overline{H}_I$ and $\overline{K}_I$ can be obtained from Eqn. (21).

Based CST and the orthogonality condition, Eqn. (37) can be simplified to

$$\begin{bmatrix} m_1 & 0 & 0 & 0 \\ 0 & m_1 & 0 & 0 \\ 0 & 0 & \text{O} & 0 \\ 0 & 0 & 0 & m_{3N^2} \end{bmatrix} \begin{Bmatrix} q_1'' \\ q_2'' \\ M \\ q_{3N^2}'' \end{Bmatrix} +$$

$$\left(\begin{bmatrix} k_1 & 0 & 0 & 0 \\ 0 & k_2 & 0 & 0 \\ 0 & 0 & \text{O} & 0 \\ 0 & 0 & 0 & k_{3N^2} \end{bmatrix} - N_d \cos Pt \begin{bmatrix} \overline{Q}_1 & 0 & 0 & 0 \\ 0 & \overline{Q}_2 & 0 & 0 \\ 0 & 0 & \text{O} & 0 \\ 0 & 0 & 0 & \overline{Q}_{3N^2} \end{bmatrix} \right) \begin{Bmatrix} q_1 \\ q_2 \\ M \\ q_{3N^2} \end{Bmatrix} = \begin{Bmatrix} 0 \\ 0 \\ M \\ 0 \end{Bmatrix} \tag{40}$$

where

$$m_I = \frac{\pi l}{2}\left[\left(I_1\overline{B}_I\right)^2 + \left(I_1\overline{A}_I\right)^2 + \left(I_1\overline{C}_I\right)^2 \right] \qquad k_I = \omega_I^2 m_I,$$

$$\overline{Q}_I = -\frac{\pi L}{2}\lambda_m^2\left[\overline{B}_I\left(I_1\overline{B}_I\right) + I_1\overline{C}_I^{\,2}\right] \quad I = 1,2,3,...,3N^2 \tag{41}$$

The coefficients $\overline{A}_I$, $\overline{B}_I$, $\overline{C}_I$ in Eqn. (40) can be obtained from Eqn. (24).

Eqns. (37) and (40) are in the form of second order differential equations with periodic coefficients of the Mathieu-Hill type. Using the method presented by Bolotin [9], the regions of unstable solutions are separated by periodic solutions. As a first approximation, the periodic solutions with period 2T can be sought in the form

$$q_I = a_I \sin\frac{Pt}{2} + b_I \cos\frac{Pt}{2} \tag{42}$$

where a_I and b_I are arbitrary constants.

By substituting Eqn. (42) into Eqns. (37) and (40) and equating the coefficients of the $\sin Pt/2$ and $\cos Pt/2$ terms, a set of linear homogeneous algebraic equations in terms of a_I and b_I can be obtained. The conditions for non-trivial solutions for the linear homogeneous algebraic equations are

$$-\frac{1}{4}\overline{P}_I^2 m_I + k_I - \frac{1}{2}N_d\overline{Q}_I = 0 \tag{43}$$

$$-\frac{1}{4}\overline{P}_I^2 m_I + k_I + \frac{1}{2}N_d\overline{Q}_I = 0 \tag{44}$$

Each unstable region is bounded by two lines which originate from a common point from the P-axis. The branches emanate at $N_d = 0$ from $2\omega_I$. The left and right branch correspond with

$$\overline{P}_I = \sqrt{\frac{4k_I + 2N_d\overline{Q}_I}{m_I}} \quad and \quad \overline{P}_I = \sqrt{\frac{4k_I - 2N_d\overline{Q}_I}{m_I}} \quad (N_d > 0)$$

or

$$\overline{P}_I = \sqrt{\frac{4k_I - 2N_d\overline{Q}_I}{m_I}} \quad and \quad \overline{P}_I = \sqrt{\frac{4k_I + 2N_d\overline{Q}_I}{m_I}} \quad (N_d < 0)$$

3. Numerical Results and Discussions

The ceramic material used in this study is silicon nitride and the metal material used is nickel [10]. The density of silicon nitride is ρ_c=2370 kg/m^{3}and that of nickel is ρ_m=8900 kg/m^3, while the Poisson's ratio is ν_c=0.24 for silicon nitride and ν_m=0.31 for nickel. These are independent of temperature. The elastic moduli are temperature-dependent and are obtained from Ng et al. [10] as

$$E_c = 348.43 \times 10^9 \left(1 - 3.070 \times 10^{-4} T + 2.160 \times 10^{-7} T^2 - 8.946 \times 10^{-11} T^3\right)$$

$$E_m = 223.95 \times 10^9 \left(1 - 2.794 \times 10^{-4} T - 3.998 \times 10^{-9} T^2\right)$$

where E_c and E_m are the elastic moduli (in Pa) of silicon nitride and nickel, respectively, and T is the temperature (in Kelvin). The elastic moduli E_c and E_m are substituted into F_c and F_m in Eqn. (3) respectively in order to compute the effective elastic moduli of the FGM shell.

Comparison Studies

Example. Based on CST, the point of origin, $P_1 = 2 \times \omega_1 \times \alpha$ [the nondimensionalized coefficient $\alpha = 2\pi R \times \sqrt{I_1 / A_{11}}$, $I = 1$, see Eqn. (39)] is presented in Table 1 for a silicon nitride-nickel FGM cylindrical shell with simply supported ends under axial extensional loading. The computation parameters are taken as $m = n = 1$, $N_0 = 0.5 N_{cr}$, $L/R = 1$, $R/h = 100$.

Table 1. Comparison of the point of origin P_1 for a simply supported silicon nitride-nickel FGM cylindrical shell under axial extensional loading

Φ	P_1 (Type A)		P_1 (Type B)	
	Present	Ng et al.[10]	Present	Ng et al.[10]
0	10.955	10.956	10.778	10.774
0.5	10.896	10.894	10.849	10.849
1.0	10.867	10.865	10.883	10.883
5.0	10.809	10.805	10.936	10.937

Table 1. (Continued)

Φ	P_1 (Type A)		P_1 (Type B)	
	Present	Ng et al.[10]	Present	Ng et al.[10]
10	10.795	10.791	10.945	10.946
∞	10.778	10.773	10.955	10.946

The results in Table 1 present the transverse modes corresponding to the point of origin P_1 , and are in a good agreement with Ng et al. [10] .

Parametric Resonance Results

The dynamic instability regions for the first order parametric resonances of a silicon nitride-nickel FGM (Type A) cylindrical shell with simply supported ends under combined static and periodic axial loads are presented in Figure 1 using CST and FSDT. The effect of FSDT on the dynamic instability regions is related to the points of origin $P_1 = 2\times\omega_6\times\alpha\ (I=6)$, $P_2 = 2\times\omega_{16}\times\alpha\ (I=16)$, and $P_3 = 2\times\omega_1\times\alpha\ (I=1)$ [see Eqn. (39), FSDT]. For the points P_1 and P_2 , the dynamic instability regions obtained from FSDT are less than those obtained from the CST, not considering shear deformation and rotary inertias. For the point P_3, the dynamic instability regions are very close for CST and FSDT, where ω_6 , ω_{16} and ω_1 denote the three lowest natural frequency using FSDT.

Figure 2 shows the effect of thickness to radius ratio (h/R) on the first unstable region (corresponding to the point of origin P_1) for a silicon nitride-nickel FGM cylindrical shell under combined static axial extensional loading and periodic axial load based on FSDT. It is observed that the points of origin of the unstable region are lower for the thinner shells. The angle φ gives a good measure of the size of the unstable region in Figure 2. Here, the unstable regions increase with increasing thicknesses.

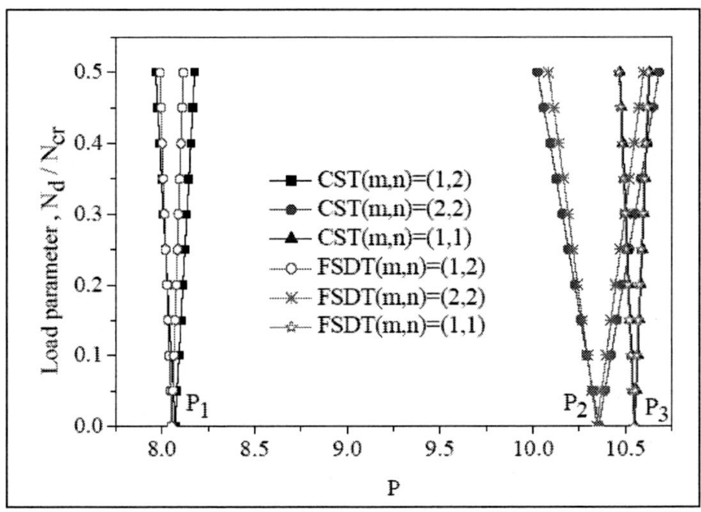

Figure 1. Comparison of classical shell theory and first-order shear deformation theory unstable regions for a simply supported silicon nitride-nickel FGM Type A cylindrical shell under combined static axial compressive loading and periodic axial loading (m=1,2 , n=1,2 , $N_0 = 0.5N_{cr}$, L/R=1.0 , T=300K, Φ=¡ h/R=0.01).

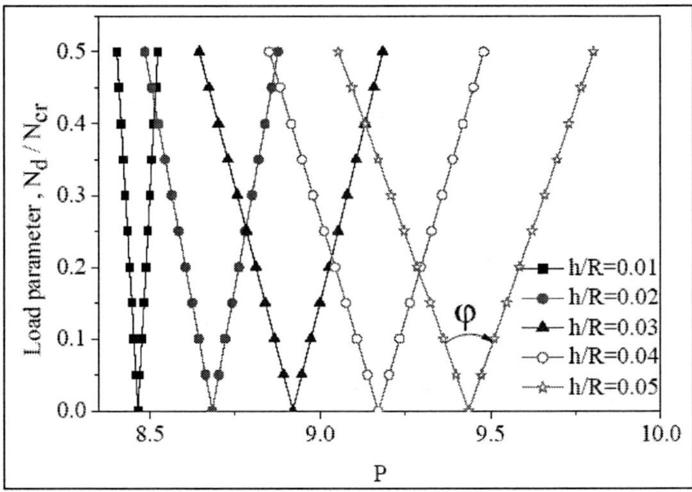

Figure 2. Effect of thickness to radius ratio *h/R* on the first unstable region (corresponding to the points of origin 1 P_1) for a simply supported silicon nitride-nickel FGM Type A cylindrical shell under combined static axial extensional loading and periodic axial load (*m*=1,2 , *n*=1,2 , $N_0 = 0.5N_{cr}$, *L/R*=1.0 , *T*=300K, Φ=1.0).

The effect of static axial compressive loading N_o/N_{cr} on unstable angles φ can be seen from the results presented in Figure3 using CST and FSDT. The points of origin are, respectively, $P_1 = 2\times\omega_6\times\alpha\ (I=6)$, $P_2 = 2\times\omega_{16}\times\alpha\ (I=16)$, $P_3 = 2\times\omega_1\times\alpha\ (I=1)$,and $P_4 = 2\times\omega_{11}\times\alpha\ (I=11)$ [see Eqn. (39), FSDT]. The first four unstable angles φ (corresponding to the points of origin P_1 , P_2 , P_3 , P_4) for a silicon nitride-nickel FGM cylindrical shell under combined static axial compressive loading and periodic axial loading are described in Figure 3(a) and Figure 3(b), respectively. The unstable region increases as the static axial load increases, and the effect of shear deformation and rotary inertias on the unstable region is dependent on the points of origin.

It is shown from the results presented in Figs. 3(a) and (b) that the unstable angles for the points P_1 , P_2 and P_4 obtained using FSDT are less than those obtained using CST, and when the static axial compressive loading N_o/N_{cr} is larger than 0.4, the third unstable angle (corresponding to the point of origin P_3 and m=1, n=1) is almost the same using CST and FSDT.

Figs. 4(a-c) show the effect of the volume fraction exponent on the unstable angle φ of a silicon nitride-nickel FGM cylindrical shell under combined static axial compressive loading and periodic axial loading based on CST and FSDT. The first three unstable regions (corresponding to the points of origin P_1 , P_2 and P_3) are described in Figs. 4(a-c), respectively. It is observed that the unstable angle φ nonlinearly increases as the volume fraction exponent Φ increases, the effect of shear deformation and rotary inertias on the unstable angles (corresponding to m=1, n=1) is not only dependent on the points of origin but also on the volume fraction exponent Φ , and the unstable angles for the point P_2 obtained by using CST and FSDT are the same.

CONCLUSIONS

This chapter reports the result of an investigation into the effect of the first-order shear deformation theory considering rotary inertia and the transverse shear strains on the dynamic instability of functionally graded cylindrical shells with simply-supported ends, under combined static and periodic axial forces.

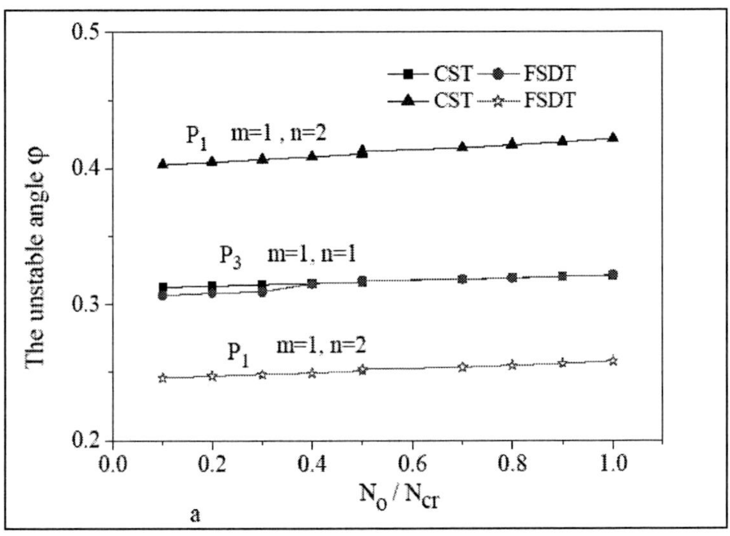

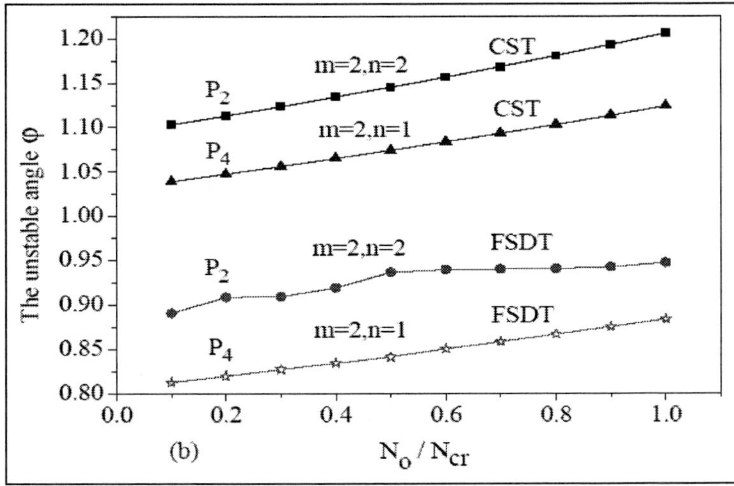

Figure 3. Unstable region φ versus axial compressive loading N_0/N_{cr} for a simply supported silicon nitride-nickel FGM Type A cylindrical shell under combined static axial compressive loading and periodic axial loading (m=1,2 , n=1,2 , L/R=1.0 , T=300K, Φ=ịh/R=0.0ị).

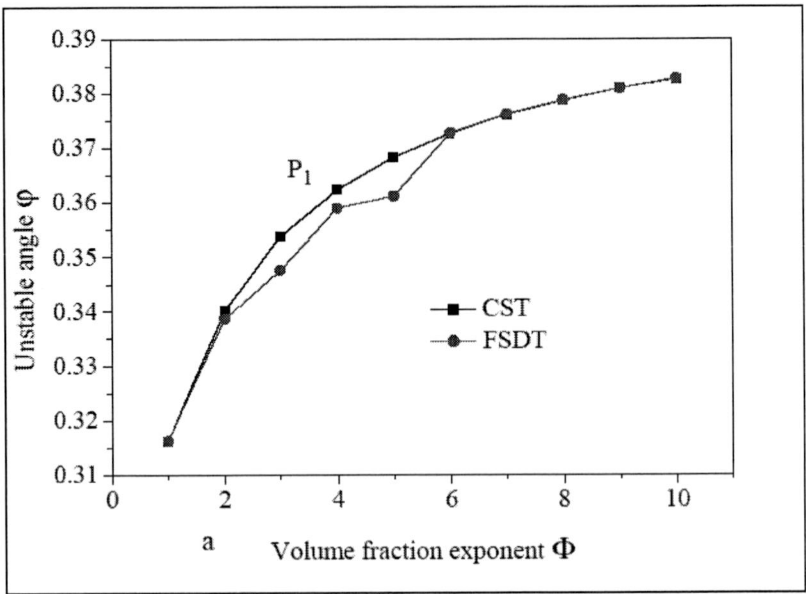
0.39
0.38
0.37
0.36
0.35
0.34
0.33
0.32
0.31
Unstable angle φ
P1
CST
FSDT
0
2
4
6
8
10
a
Volume fraction exponent Φ

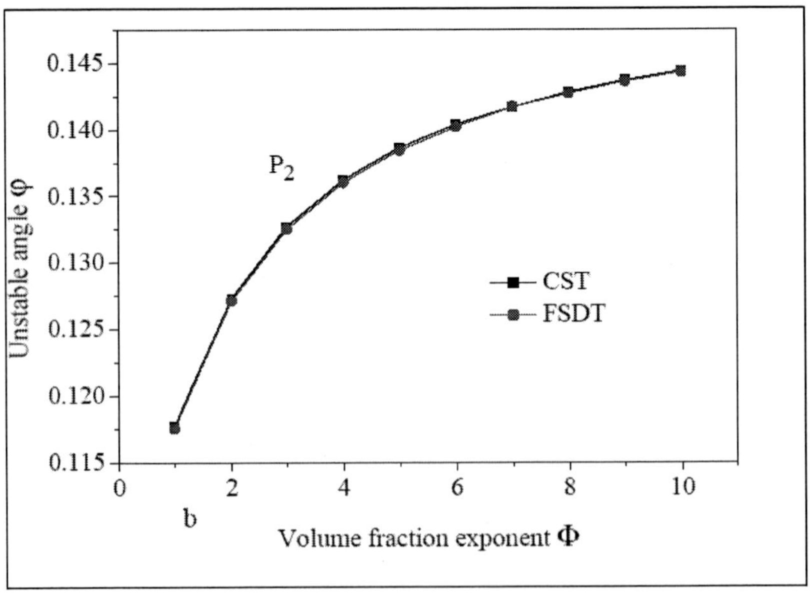
0.145
0.140
0.135
0.130
0.125
0.120
0.115
Unstable angle φ
P2
CST
FSDT
0
2
4
6
8
10
b
Volume fraction exponent Φ

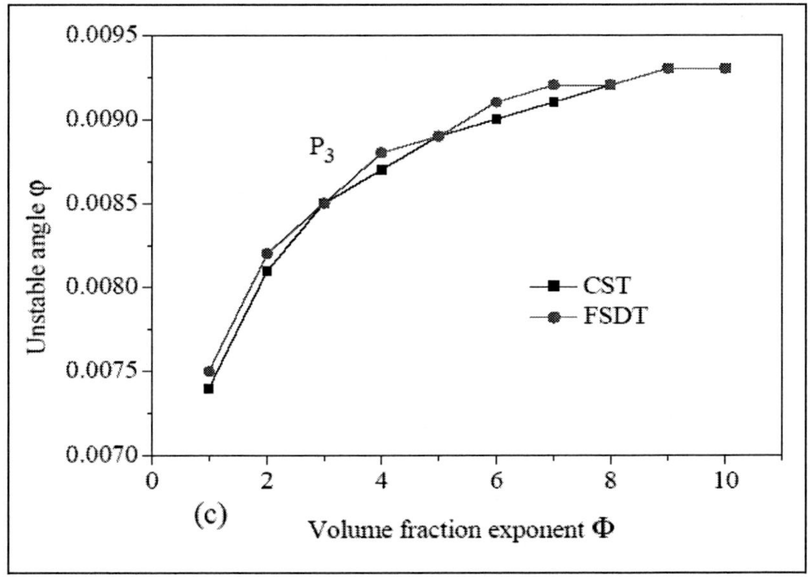

Figure 4. Effect of the volume fraction exponent Φ on the unstable angles φ for a simply supported silicon nitride-nickel FGM Type A cylindrical shell under combined static axial compressive loading and periodic axial loading (m=1 , n=1 , $N_0 = 0.5N_{cr}$, L/R=1.0 , T=300K, Φ=1.0 , h/R=0.01).

The result obtained using the first-order shear deformation theory (FSDT) is compared with that obtained using classical shell theory (CST). The differences between the results from FSDT and the results from CST increase as the deformation mode and thickness increase. It was found that reasonable control can be achieved on the dynamic instability regions by correctly varying the ratio of length to radius, the ratio of thickness to radius, the amplitude of static axial load, thermal environment, and the volume fraction exponent.

REFERENCES

[1] M. Koizumi and M. Niino, Overview of FGM research in Japan, MRS Bull. 20 (1995) 19– 21.

[2] J. N. Reddy and C. D. Chin, Thermomechanical Analysis of Functionally Graded Cylinders and Plates, *J. Thermal Stresses.* 26(1) (1998) 593-626.

[3] A.H. Sofiyev, The stability of compositionally graded ceramic–metal cylindrical shells under a periodic axial impulsive loading, *Compos. Struct.* 69 (2005) 247–257.

[4] M. Darabi, M. Darvizeh and A. Darvizeh, Non-linear analysis of dynamic stability for functionally graded cylindrical shells under periodic axial loading, *Compos. Struct.* 83(2) (2008) 201-211.

[5] H. Matsunaga, Free vibration and stability of functionally graded circular cylindrical shells according to a 2D higher order deformation theory, *Compos. Struct.* 88(4) (2009) 519-531.

[6] H. Matsunaga, Free vibration and stability of functionally graded shallow shells according to a 2D higher-order deformation theory, *Compos. Struct.* 84(2) (2008)132-146.

[7] S.P. Timoshenko and S. Woinowsky-Krieger, Theory of Plates and Shells, 2nd Edition, McGraw-Hill, New York, (1959).

[8] J. N. Reddy, Mechanics of Laminated Composite Plates and Shells, Second Edition, CRC Press, New York, (2004).

[9] V. V. Bolotin, The Dynamic Stability of Elastic Systems, Holden-Day, San Francisco, (1964).

[10] T.Y. Ng, K.M Lam., K.M. Liew and J. N. Reddy, Dynamic stability analysis of functionally graded cylindrical shells under periodic axial loading, *Int. J. Solid Struct.* 38 (2001) 1295-309.

Chapter 3

Vibration and Buckling Analysis of Two-Layered Functionally Graded Cylindrical Shells

Abstract

This research investigates the free vibration and buckling of a two-layered cylindrical shell made of inner functionally graded (FG) and outer isotropic elastic layer, subjected to combined static and periodic axial forces. Material properties of functionally graded cylindrical shell are considered as temperature dependent and graded in the thickness direction according to a power-law distribution in terms of the volume fractions of the constituents. Theoretical formulations are presented based on two different methods of first-order shear deformation theory (FSDT) considering the transverse shear strains and the rotary inertias and the classical shell theory (CST). The results obtained show that the transverse shear and rotary inertias have considerable effect on the fundamental frequency of the FG cylindrical shell. The results for nondimensional natural frequency are in a close agreement with those in literature. It is inferred from the results that the geometry parameters and material composition of the shell have significant effect on the critical axial force, so that the minimum critical load is obtained for fully metal shell. Good agreement between theoretical and finite element results validates the approach. It is concluded that the presence of an additional elastic layer significantly increases the nondimensional natural frequency, the buckling resistance and hence the elastic stability in axial compression with respect to a FG hollow cylinder.

1. INTRODUCTION

Functionally graded materials (FGMs) are now developed for general use as structural components in extremely high temperature environments. Unlike fiber-matrix composites which have a mismatch of mechanical properties across an interface of two discrete materials bonded together and may result in debonding at high temperatures, FGMs have the advantage of being able to withstand high temperature environments while maintaining their structural integrity, as presented at the works done by the authors, in which the behavior of FGM plates combined with smart structures are investigated [1-2].

Shen [3] showed that in shell buckling, there is a boundary layer phenomenon where prebuckling and buckling displacement vary rapidly and performed the postbuckling analysis for imperfect, stiffened and multilayered cylindrical shells [4].

Cylindrical shells in engineering structures with large aspect ratios are typically stiffened against buckling by circumferential and longitudinal members, known as ring and stringers stiffeners, respectively. Accordingly, sandwich construction with lightweight additional layer materials has well known advantages of stiffness and strength to weight. These structures are widely used in applications where weight consideration has great importance. Cylindrical shell with an elastic layer under axial pressure is well known to be highly efficient in terms of combining high stiffness to weight. These structures consist of a fully condensed shell material supported by a low density external/internal layer. The second author analyzed the elastic buckling of a thin cylindrical shell supported by an elastic core [5] showing that this structural configuration achieves significant weight saving compared to a hollow cylinder. Hutchinson et al. [6] studied buckling of cylindrical shells with metal foam cores subjected to axial compression. They obtained optimal outer shell thickness, core thickness and core density by minimizing the weight of geometrically perfect shell with a specified load carrying capacity. Agrawal et al. [7] carried out a similar research in which the weight compressions of optimized stiffened, unstiffened, and sandwich cylindrical shells are investigated.

Although the two-layered cylindrical shells have been taken up by many aforementioned researchers, but to the authors' best knowledge few researches dealing with the modeling of FGM cylindrical shells embedded with isotropic elastic layers have been reported in literature, thus motivating the authors to study the free vibration and buckling behavior of two-layered Functionally Graded cylindrical shells.

Two-layered FG cylindrical shells have many applications. In the nuclear reactors, 316L stainless steel is used as elastic main body and functionally graded material used as inner anti-decay/anti-radiation/high temperature resistant layer, making it helpful to solve the problem of unmatched thermal expansion coefficients and elastic modulus among different materials. In this regard B4C/Cu and SiC/Cu, and tungsten based alloys were introduced as Plasma Facing Materials by Xiang et al. [8] and Baluc et al.[9], respectively.

Other feature of application of two-layered FG cylindrical shells, is active vibration control of composite laminated cylindrical shells via surface-bonded magnetostrictive or piezoelectric layers [10]. Smart materials are widely used in modern engineering due to their direct and inverse effects. The use of a piezoelectric layer as distributed sensors and actuators in active structure control such as noise attenuation, deformation control and vibration suppression have attracted serious attention. On the other hand, in FGM the possibility of tailoring the desired thermo-mechanical properties holds enormous application potential for it. Therefore the integration of piezoelectric materials and composite materials or FGM has become the subject of focus in the area of smart materials and structures and numerous papers on this subject have been published [11].

This chapter studies the free vibration and buckling of a two-layered FG cylindrical shell subjected to combined static and periodic axial forces. The model consists of one isotropic elastic cylindrical shell with functionally graded cylindrical core. A comparison is made to show the free vibration behavior of the structure through results obtained by using two different methods such as the first-order shear deformation theory considering the transverse shear strains and the rotary inertias and the classical shell theory. Also, the results for natural frequency are compared between two-layered FG cylindrical shell and hollow FGM cylindrical shell. The elastic buckling of the two-layered FG cylindrical shell has been analyzed, and its buckling resistance was compared with that of a FG hollow cylindrical shell. Material properties of functionally graded cylindrical shell are considered as temperature dependent and graded in the thickness direction according to a power-law distribution in terms of the volume fractions of the constituents. It is concluded that the presence of an additional elastic layer significantly increases the nondimensional natural frequency, the buckling resistance and so the elastic stability in axial compression with respect to a hollow FG cylinder.

2. Modeling of the Problem

A Functionally Graded, closed-ended, perfectly circular cylindrical shell embedded with an isotropic elastic layer with simply supported end conditions has been considered in this research. The FG shell has a uniform thickness *h*, radius *R*, length *L*, density $\rho_{eff}(T,z)$, Young's modulus $E_{eff}(T,z)$ and Poisson ratio $\upsilon_{eff}(T,z)$. The properties are assumed to be graded in the thickness direction according to a power-law distribution in terms of the volume fractions of the constituents. The isotropic elastic layer's properties are: density ρ_e, Young's modulus E_e, Poisson ratio υ_e, and a uniform thickness *t* as shown in Figure 1. Coordinates x, θ, and z axes are defined with corresponding displacements u, v and w.

The material properties of FG layer with both temperature and position dependent are accurately modeled, using a simple rule of mixtures for the stiffness parameters coupled with the temperature dependency properties of the constituents. The volume fraction is described by a spatial function as

$$V_c(z) = (z/h + 1/2)^g \ , \ g \geq 0 \tag{1}$$

where subscript *c* refers to the material *c* and *g* expresses the volume fraction exponent, called FGM index factor. The combination of these functions gives rise to the effective properties of functionally graded materials. The corresponding effective material properties for the FG cylindrical shell that is ceramic rich at the inner surface and metal rich at the outer surface are expressed as

$$F_{eff}(T,z) = F_m(T)V_c(z) + F_c(T)(1 - V_c(z)) \tag{2}$$

Here subscript *c* refers to the ceramic, subscript *m* refers to the metal and F_{eff} is the effective material property of the FGM cylindrical shell, including the effective elastic modulus, effective mass density and effective Poisson's ratio. F_c and F_m are the temperature dependent properties of the ceramic and metal, respectively. Thus the effective material properties are given by

$$E_{eff}(T,z) = (E_m(T) - E_c(T))(z/h + 1/2)^g + E_c(T)$$
$$\rho_{eff}(T,z) = (\rho_m(T) - \rho_c(T))(z/h + 1/2)^g + \rho_c(T)$$

$$v_{eff}(T,z) = (v_m(T) - v_c(T))(z/h + 1/2)^g + v_c(T) \tag{3}$$

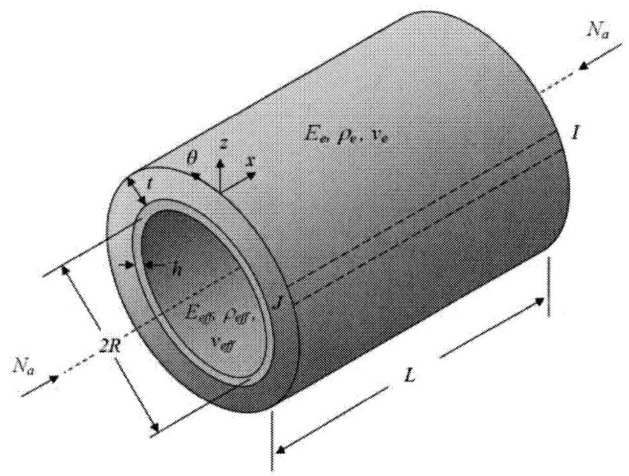

Figure 1: Geometry of a two-layered functionally graded cylindrical shell with loading configuration.

Assuming that the shell thickness is much smaller than its radius, the spring constant k_e can be found from the result of stress in the z direction for a flat strip element (I J in Figure 1) with an elastic foundation subjected to a sinusoidal displacement in the z direction [12]:

$$\sigma_z = -\frac{2\pi E_e}{(3 - \upsilon_e)(1 + \upsilon_e)} \times \frac{m}{L} w \tag{4}$$

where σ_z, *m* and *L* are stress in the *z* direction, longitudinal wave number and shell length, respectively. For an FG cylindrical shell embedded with an elastic layer under external pressure *q*, we have

$$q = \sigma_z = -k_e w \tag{5}$$

in which *w* is the radial displacement of the shell. Equations (4) and (5) lead to

$$k_e = -\frac{2E_e}{(3-\upsilon_e)(1+\upsilon_e)} \times \frac{1}{\lambda} \tag{6}$$

where λ is buckling wavelength parameter

$$\lambda = \frac{l}{m\pi} \tag{7}$$

Assuming the shell under axial compression, the pulsating axial load is given by

$$N_a = N_0 + N_d \cos Pt \tag{8}$$

where P is the frequency of excitation in radians per unit time.

3. THEORETICAL FORMULATIONS

Based on the first-order shear deformation theory, the equations of motion for an FG cylindrical shell with external isotropic elastic layer under axially dynamic load are as follows [13]

$$\frac{\partial N_x}{\partial x} + \frac{1}{R}\frac{\partial N_{x\theta}}{\partial \theta} = I_1 \ddot{u} + I_2 \ddot{\Phi}_x \tag{9-a}$$

$$\frac{\partial N_{x\theta}}{\partial x} + \frac{1}{R}\frac{\partial N_\theta}{\partial \theta} + \frac{1}{R} Q_\theta + N_a \frac{\partial^2 v}{\partial x^2} = I_1 \ddot{v} + I_2 \ddot{\Phi}_\theta \tag{9-b}$$

$$\frac{\partial Q_x}{\partial x} + \frac{1}{R}\frac{\partial Q_\theta}{\partial \theta} - \frac{1}{R} N_\theta + N_a \frac{\partial^2 w}{\partial x^2} + k_e w = I_1 \ddot{w} \tag{9-c}$$

$$\frac{\partial M_x}{\partial x} + \frac{1}{R}\frac{\partial M_{x\theta}}{\partial \theta} - Q_x = I_2 \ddot{u} + I_3 \ddot{\Phi}_x \tag{9-d}$$

$$\frac{\partial M_{x\theta}}{\partial x}+\frac{1}{R}\frac{\partial M_{\theta}}{\partial \theta}-Q_{\theta}=I_2\ddot{v}+I_3\ddot{\Phi}_{\theta} \tag{9-e}$$

where Φ_1 and Φ_2 are the rotations of a normal to the reference surface and I_i (i = 1,2,3) is the mass inertia terms defined as

$$I_i=\int_{-h/2}^{h/2} z^i\,\rho(z)dz \tag{10}$$

in which $\rho(z)$ is the effective mass density of functionally graded materials The stress resultants of FG cylindrical shell are given by

$$\begin{pmatrix} N_x \\ N_\theta \\ N_{x\theta} \\ M_x \\ M_\theta \\ M_{x\theta} \end{pmatrix}=\begin{pmatrix} A_{11} & A_{12} & 0 & B_{11} & B_{12} & 0 \\ A_{12} & A_{22} & 0 & B_{12} & B_{22} & 0 \\ 0 & 0 & A_{66} & 0 & 0 & B_{66} \\ B_{11} & B_{12} & 0 & D_{11} & D_{12} & 0 \\ B_{12} & B_{22} & 0 & D_{12} & D_{22} & 0 \\ 0 & 0 & B_{66} & 0 & 0 & D_{66} \end{pmatrix}\begin{pmatrix} \varepsilon_x \\ \varepsilon_\theta \\ \varepsilon_{x\theta} \\ \kappa_x \\ \kappa_\theta \\ \kappa_{x\theta} \end{pmatrix}$$

$$\begin{pmatrix} Q_x \\ Q_\theta \end{pmatrix}=\begin{pmatrix} C_{44} & 0 \\ 0 & C_{55} \end{pmatrix}\begin{pmatrix} \varepsilon_{xz} \\ \varepsilon_{\theta z} \end{pmatrix} \tag{11}$$

where A_{ij} , B_{ij} , D_{ij} and C_{ij} are ,respectively, the extensional, coupling, bending, and shear stiffness, given by

$$\left(A_{ij},B_{ij},D_{ij}\right)=\int_{-h/2}^{h/2} Q_{ij}\,\left(1,z,z^2\right)dz \quad , \quad (i=1,2,6)$$

$$Q_{11}=Q_{22}=\frac{E_{eff}}{1-\nu_{eff}^2}$$

$$Q_{12}=Q_{12}=\frac{\nu_{eff}\,E_{eff}}{A\left(1-\nu_{eff}^2\right)},\ Q_{66}=\frac{E_{eff}}{2A\left(1+\nu_{eff}\right)}$$

$$Q_{44}=\kappa\frac{E_{eff}}{2\left(1+\nu_{eff}\right)},\ Q_{55}=Q_{44}/A,\ A=1+z/R$$

$$C_{44} = \int_{-h/2}^{h/2} Q_{44}\ dz\,,\ C_{55} = \int_{-h/2}^{h/2} Q_{55}\ dz \tag{12}$$

where E_{eff} and ν_{eff} are the effective elastic modulus and effective Poisson's ratio of FGM, respectively. κ is the shear correction factor introduced by Reddy [**14**] and is equal to 5/6 . The strains are expressed as:

$$\varepsilon_x = \frac{\partial u}{\partial x} \quad , \quad \varepsilon_\theta = \frac{1}{R}\left(\frac{\partial v}{\partial \theta} + w\right) \quad , \quad \varepsilon_{x\theta} = \frac{\partial v}{\partial x} + \frac{1}{R}\frac{\partial u}{\partial \theta} \quad , \quad \varepsilon_{xz} = \phi_\theta + \frac{1}{R}\frac{\partial w}{\partial \theta}$$
$$\varepsilon_{\theta z} = \phi_x + \frac{\partial w}{\partial x} \quad , \quad \kappa_x = \frac{\partial \phi_x}{\partial x} \quad , \quad \kappa_\theta = \frac{1}{R}\frac{\partial \phi_\theta}{\partial \theta} \quad , \quad \varepsilon_{x\theta} = \frac{\partial \phi_\theta}{\partial x} + \frac{1}{R}\frac{\partial \phi_x}{\partial \theta} \tag{13}$$

Utilizing Eqs. (9), (11) and (13), the equations of motion can be expressed in terms of generalized displacement (u,v,w,Φ_1, Φ_2) as follows

$$L_1(u, v, w, \phi_x, \phi_\theta) = I_1\ddot{u} + I_2\ddot{\phi}_x \tag{14}$$

$$L_2(u, v, w, \phi_x, \phi_\theta) + N_a\frac{\partial^2 v}{\partial x^2} = I_1\ddot{v} + I_2\ddot{\phi}_\theta \tag{15}$$

$$L_3(u, v, w, \phi_x, \phi_\theta) + N_a\frac{\partial^2 w}{\partial x^2} + k_e w = I_1\ddot{w} \tag{16}$$

$$L_4(u, v, w, \phi_x, \phi_\theta) = I_2\ddot{u} + I_3\ddot{\phi}_x \tag{17}$$

$$L_5(u, v, w, \phi_x, \phi_\theta) = I_2\ddot{v} + I_3\ddot{\phi}_\theta \tag{18}$$

By neglecting the terms I_1 and I_3 involving in equations (9) and setting

$$\phi_x = -\frac{\partial w}{\partial x} \quad , \quad \phi_\theta = -\frac{1}{R}\frac{\partial w}{\partial \theta} \tag{19}$$

the equations of motion based on a classical shell theory can be easily obtained. Here, the two ends of cylindrical shell are considered as simply supported, so that a solution for the motion equations (14)-(18) can be described by

$$
\begin{aligned}
u_{mn} &= \bar{A}_{mn} e^{i\omega t} \cos\lambda_m x \cos n\theta \\
v_{mn} &= \bar{B}_{mn} e^{i\omega t} \sin\lambda_m x \sin n\theta \\
w_{mn} &= \bar{C}_{mn} e^{i\omega t} \sin\lambda_m x \cos n\theta \\
\phi_{xmn} &= \bar{H}_{mn} e^{i\omega t} \cos\lambda_m x \cos n\theta \\
\phi_{\theta mn} &= \bar{K}_{mn} e^{i\omega t} \sin\lambda_m x \sin n\theta
\end{aligned} \qquad (20)
$$

where $\lambda_m = \frac{m\pi}{L}$, n represents the number of circumferential waves and m represents the number of axial half-waves. Substituting Eq. (20) into Eqs.(14)-(18) and letting $N_d = 0$ in Eq.(8), yields

$$
\left(\begin{bmatrix} T_{11} & T_{12} & T_{13} & T_{14} & T_{15} \\ T_{21} & T_{22} + \lambda_m^2 N_0 & T_{23} & T_{24} & T_{25} \\ T_{31} & T_{32} & T_{33} + \lambda_m^2 N_0 & T_{34} & T_{35} \\ T_{41} & T_{42} & T_{43} & T_{44} & T_{45} \\ T_{51} & T_{52} & T_{53} & T_{54} & T_{55} \end{bmatrix} - \omega^2 \begin{bmatrix} I_1 & 0 & 0 & I_2 & 0 \\ 0 & I_1 & 0 & 0 & I_2 \\ 0 & 0 & I_1 & 0 & 0 \\ I_2 & 0 & 0 & I_3 & 0 \\ 0 & I_2 & 0 & 0 & I_3 \end{bmatrix} \right) \begin{pmatrix} \bar{A}_{mn} \\ \bar{B}_{mn} \\ \bar{C}_{mn} \\ \bar{H}_{mn} \\ \bar{K}_{mn} \end{pmatrix} = \begin{pmatrix} 0 \\ 0 \\ 0 \\ 0 \\ 0 \end{pmatrix} \qquad (21)
$$

where T_{ij}'s are

$$
\begin{aligned}
T_{11} &= A_{11}\lambda_m^2 + A_{66}\bar{n}^2 \\
T_{22} &= A_{66}\lambda_m^2 + A_{22}\bar{n}^2 + \frac{C_{55}}{R^2} T_{33} = C_{44}\lambda_m^2 + C_{55}\bar{n}^2 + \frac{A_{22}}{R^2} - k_e
\end{aligned} \qquad (22)
$$

$$
\begin{aligned}
T_{44} &= D_{11}\lambda_m^2 + D_{66}\bar{n}^2 + C_{44} \\
T_{55} &= D_{66}\lambda_m^2 + D_{22}\bar{n}^2 + C_{55} \\
T_{12} &= T_{21} = -(A_{12} + A_{66})\lambda_m \bar{n}^2 \\
T_{13} &+ T_{31} = -A_{12}\lambda_m \bar{n} T_{14} = T_{41} = B_{11}\lambda_m^2 + B_{66}\bar{n}^2 \\
T_{15} &= T_{51} = -(B_{12} + B_{66})\lambda_m \bar{n}^2 \\
T_{23} &= T_{32} = (A_{22} + C_{55})\bar{n} \\
T_{24} &= T_{42} = -(B_{12} + B_{66})\lambda_m \bar{n}
\end{aligned}
$$

$$T_{25} = T_{52} = B_{66}\lambda_m^2 + B_{22}\bar{n}^2 - \frac{C_{55}}{R^2}$$

$$T_{34} = T_{43} = \left(C_{44} - \frac{B_{12}}{R}\right)\lambda_m$$

$$T_{35} = T_{53} = -\left(C_{55}\frac{B_{22}}{R}\right)\lambda_m$$

$$T_{45} = T_{54} = -(D_{12} + D_{66})\lambda_m\bar{n}$$

Here $\bar{n} = n/R$. The nontrivial solution of the eigen-equation (21) gives the natural frequencies of FGM cylindrical shell with outer elastic layer under the static axial load, and the corresponding mode shapes. From Eq. (21), the static buckling load N_{0cr} of the cylindrical shell can be easily determined by the following equation

$$\begin{vmatrix} T_{11} & T_{12} & T_{13} & T_{14} & T_{15} \\ T_{21} & T_{22} + \lambda_m^2 N_{0cr} & T_{23} & T_{24} & T_{25} \\ T_{31} & T_{32} & T_{33} + \lambda_m^2 N_{0cr} & T_{34} & T_{35} \\ T_{41} & T_{42} & T_{43} & T_{44} & T_{45} \\ T_{51} & T_{52} & T_{53} & T_{54} & T_{55} \end{vmatrix} = 0 \tag{23}$$

The static buckling equation (23) is simplified to

$$a(L,m,n)N_{0cr}^2 + b(L,m,n)N_{0cr} + c(L,m,n) = 0 \qquad m,n = 1,2,3,4,\ldots,\infty \tag{24}$$

In the above formula, the buckling mode m and n which make N_{0cr} minimum are used to determine the critical buckling load N_{0cr} of the two-layered FG cylindrical shell with given geometry parameters and material parameters. Neglecting rotary inertias and the transverse shear strains in Eqs.(21), (23) and (24), the corresponding eigen-equation and static buckling equation of the two-layered FG cylindrical shell based on a classical shell theory are expressed as

$$\left\{\begin{bmatrix} \bar{T}_{11} & \bar{T}_{12} & \bar{T}_{13} \\ \bar{T}_{21} & \bar{T}_{22} + \lambda_m^2 N_0 & \bar{T}_{23} \\ \bar{T}_{31} & \bar{T}_{32} & \bar{T}_{33} + \lambda_m^2 N_0 \end{bmatrix} - \omega^2 \begin{bmatrix} I_1 & 0 & 0 \\ 0 & I_1 & 0 \\ 0 & 0 & I_1 \end{bmatrix}\right\} \begin{Bmatrix} \bar{A}_{mn} \\ \bar{B}_{mn} \\ \bar{C}_{mn} \end{Bmatrix} = \begin{Bmatrix} 0 \\ 0 \\ 0 \end{Bmatrix} \tag{25}$$

$$\begin{vmatrix} \bar{T}_{11} & \bar{T}_{12} & \bar{T}_{13} \\ \bar{T}_{21} & \bar{T}_{22} + \lambda_m^2 N_0 & \bar{T}_{23} \\ \bar{T}_{31} & \bar{T}_{32} & \bar{T}_{33} + \lambda_m^2 N_0 \end{vmatrix} = 0 \tag{26}$$

where *Tij* 's are

$$T_{11} = A_{11}\lambda_m^2 + A_{66}\bar{n}^2 \tag{27}$$

$$T_{12} = -(A_{12} + A_{66})\lambda_m \bar{n}$$

$$T_{13} = T_{31} = -\frac{1}{R}A_{12}\lambda_m - B_{11}\lambda_m^3 - (2B_{66} + B_{12})\,\lambda_m \bar{n}^2$$

$$T_{21} = -\left(A_{12} + A_{66} + \frac{B_{12} + B_{66}}{R}\right)\lambda_m \bar{n}$$

$$T_{22} = A_{66}\lambda_m^2 + A_{22}\bar{n}^2 + \frac{1}{R}(B_{66}\lambda_m^2 + B_{22}\bar{n}^2)$$

$$T_{23} = (2B_{66} + B_{12} + 2D_{66} + D_{12})\lambda_m^2 \bar{n} + \frac{A_{22}}{R}\bar{n} - B_{22}\bar{n}^3 + \frac{B_{22}}{R^2}\bar{n} - \frac{D_{22}}{R}\bar{n}^3$$

$$T_{32} = (2B_{66} + B_{12})\lambda_m^2 \bar{n} + \frac{A_{22}}{R}\bar{n} - B_{22}\bar{n}^3$$

$$T_{33} = (4D_{66} + 2D_{12})\lambda_m^2 \bar{n}^2 + 2\frac{B_{12}}{R}\lambda_m^2 + 2\frac{B_{22}}{R}\bar{n}^2 + \frac{A_{22}}{R^2} + D_{11}\lambda_m^4 + D_{22}\bar{n}^4 - k_s$$

Thus the static buckling equation (26) is simplified to

$$\bar{a}(L,m,n)N_{0cr}^2 + \bar{b}(L,m,n)N_{0cr} + \bar{c}(L,m,n) = 0 \qquad m,n = 1,2,3,4,\ldots,\infty \tag{28}$$

4. Numerical Results and Discussions

The ceramic material used in this study is silicon nitride and the metal material used is nickel [13]. The elastic core is assumed to be from aluminum 7070-T7651. The material properties are listed in Table 1 and the elastic modulus are given by [13]

$$E_c = E_{0c}(1 - c_{1c}T + c_{2c}T^2 + c_{3c}T^3)$$
$$E_m = E_{0m}(1 - c_{1m}T + c_{2m}T^2 + c_{3m}T^3) \quad (29)$$

where E_c and E_m are the elastic modulus of silicon nitride and nickel, respectively, and T is the temperature in Kelvin. The constant coefficients can be found in Table 1. The elastic moduli E_c and E_m can be substituted into F_c and F_m in Eq.(2) respectively resulting the effective elastic moduli of the plate.

Based on the classical shell theory, the nondimensional natural frequency (λ) for the fully ceramic cylindrical shell with simply supported ends and without elastic part is obtained, which is in a close agreement with that of Radwan and Genin [15] as shown in Figure 2.

Figures 3 and 4 show the results of nondimensionalized fundamental frequencies $\Omega = 2\omega \times \alpha$; of a silicon nitride-nickel FGM cylindrical shell with simply supported ends based on two different shell theories such as the classical shell theory and the first-order shear deformation theory in which α is the nondimensionalized coefficient defined as $\alpha = 2\pi R\sqrt{I_1/A_{11}}$ and ω denotes the fundamental frequency. The problems taken up here are mostly cylindrical shells of intermediate length. For such shells of functionally graded material, the effect of transverse shear and rotary inertias on the fundamental frequency can be seen in Figs.3-4.

It is observed from Figure 3 that the effect of transverse shear and rotary inertias on the fundamental frequency of the FG cylindrical shell gradually increases with the increase of the value of n from 3 onwards. Figure 4 shows that the difference between the fundamental frequencies obtained by FSDT and CST, increases with the increase of *h/R*. This is due to the fact that the cylindrical shell has larger stiffness as the ratio *h/R* of thickness to radius increases.

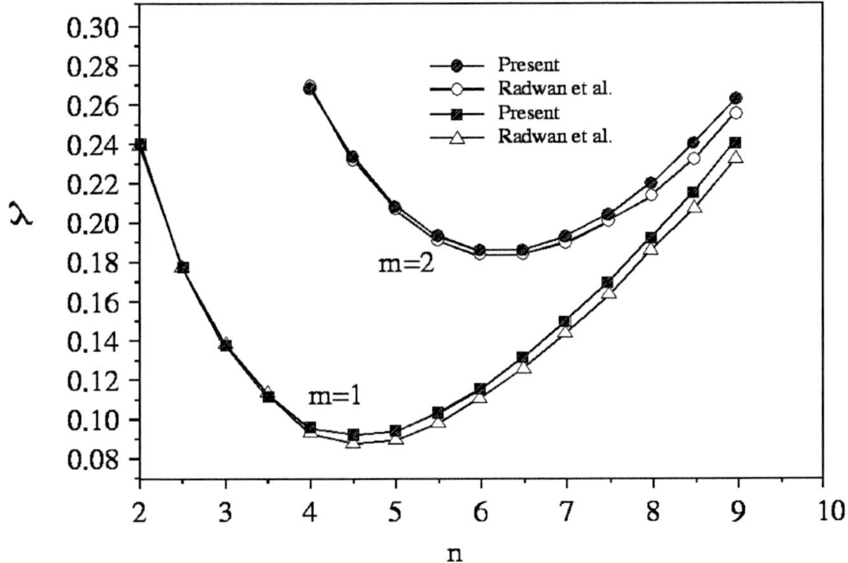

Figure 2. Non-dimensional fundamental frequency λ versus circumferential modes (m,n) for simply supported cylindrical shell. [h/R=0.01, L/R=2.5, $\lambda^2 = (I_1 R^2/E)\omega^2$], $E = E_c h/(1 - v_c^2)$, $N_0 = 0$. ω denotes the fundamental frequency.

Table 1. Material properties [13]

Density	$\rho_c = 2370\ kg/m^3$ $\rho_m = 8900\ kg/m^3$ $\rho_e = 2770\ kg/m^3$	Poisson's ratio	$\nu_c = 0.24$ $\nu_m = 0.31$ $\nu_e = 0.3$
Constant coefficients for Ceramic	$c_{1c} = 3.07 \times 10^{-4}$ $c_{2c} = 2.16 \times 10^{-7}$ $c_{3c} = 8.946 \times 10^{-11}$	Constant coefficients for Metal	$c_{1m} = 2.794 \times 10^{-4}$ $c_{2m} = -3.998 \times 10^{-9}$ $c_{3m} = 0$
Module of Elasticity	$E_{0c} = 348.42\ GPa$ $E_{0m} = 223.95\ GPa$ $E_e = 70.0\ GPa$		

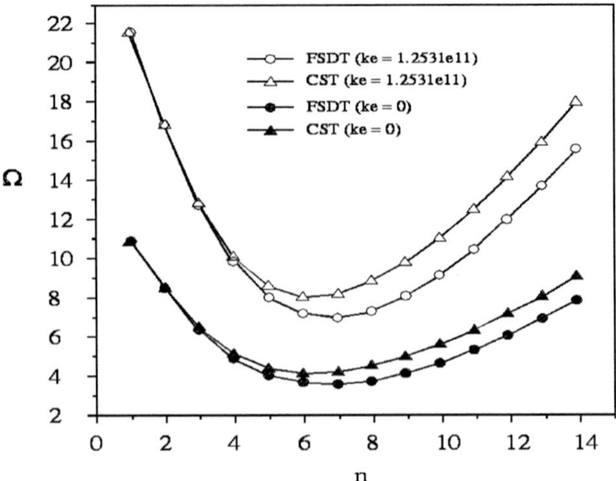

Figure 3: Non-dimensional fundamental frequency Ω ,versus circumferential mode n for a simply supported silicon nitride-nickel FG cylindrical shell under axial extensional loading, based on two different solving methods.

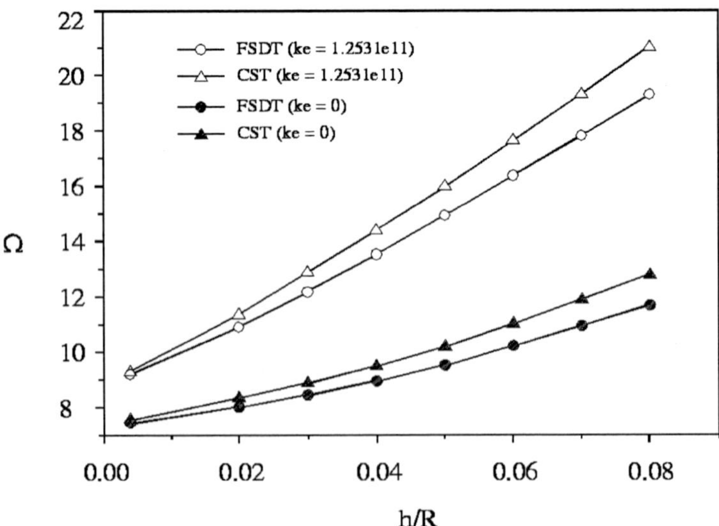

Figure 4. Non-dimensional fundamental frequency Ω ,versus thickness to radius ratio h/R for a simply supported silicon nitride-nickel FG cylindrical shell with/without elastic layer under axial extensional loading, based on two different solving methods.

It is also depicted by Figs. 3 and 4 that elastic layer increases the natural frequency of the shell. As one can observe, the presence of an elastic layer significantly increases the natural frequency with respect to a hollow FG cylinder. It is due to the fact that FG cylinder embedded with elastic layer has greater stiffness with respect to FG hollow cylinder. It means that utilizing an elastic layer increases the stability of the structure. Also, the critical force per unit circumferential length of the FG cylindrical shell with an elastic layer, N_{0cr}, versus aspect ratio *R/h* is shown in Figure 5. Results are illustrated for *g=0, 1, 3* and *200*. It is obvious that as the shell radius increases, the buckling load decreases. The variation of the critical load is more pronounced in smaller amount of aspect ratio. The effect of FGM index factor, *g*, on the critical load can be better seen in Figure 6. In addition finite element modeling is carried out to predict the behavior of two-layered FG cylindrical shell. Good agreement can be seen between theoretical and finite element results.

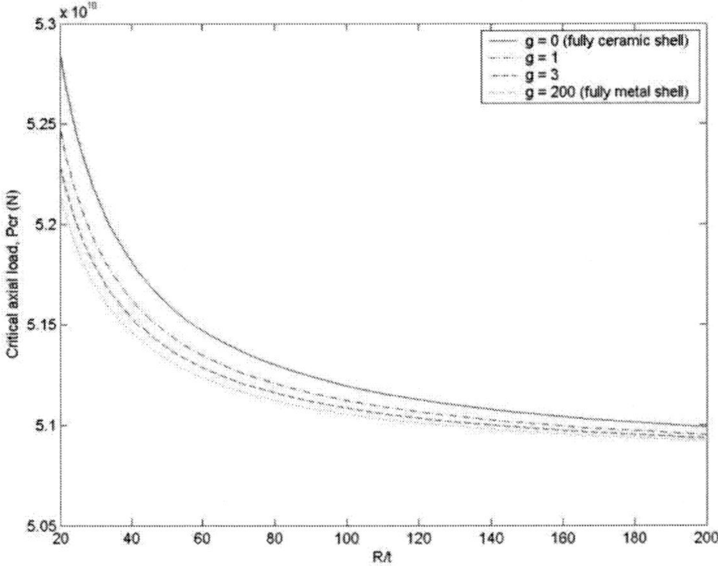

Figure 5. Critical axial force per unit circumferential length for a simply supported silicon nitride-nickel FG cylindrical shell with an elastic layer versus aspect ratio (L/R=2).

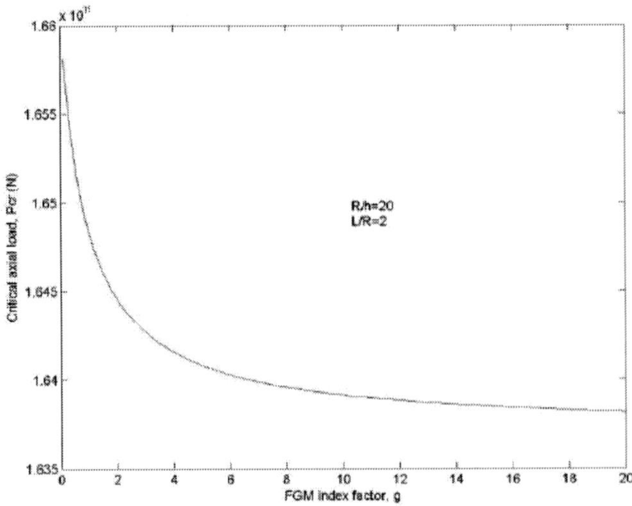

Figure 6. Critical axial force per unit circumferential length for a simply supported silicon nitride-nickel FG cylindrical shell with an elastic layer versus FGM index factor, (R/h=20, L/R=2), comparison with finite element results.

Figure 7 represents the variation of axial compressive buckling load of FG cylindrical shell with an elastic layer versus L/mR for different values of index factor, *g*. The exact relationship between N_{0cr} and L/mR is obtained via classical shell theory and for n=1. For each value of *g*, the critical load is minimum. Thus the critical load is the minimum point of the curve in Fig 7. It is also inferred that the minimum critical load can be obtained for fully metal shell, while $g = \infty$.

Figure 8 represents the effect of the elastic layer on the buckling behavior of the FG cylindrical shell. As one can observe from Figure 8, the presence of an elastic layer significantly increases the buckling resistance in axial compression with respect to a hollow FG cylinder, thus increasing the stability of the structure.

As noted before, the elastic layer has significant effect on both natural frequency and buckling resistance. Therefore it comes to the conclusion that the implication of additional elastic layers leads to stiffness and stability increasing in cylindrical shells. To show this fact, the performance of a thin walled FG cylindrical shell with a compliant elastic layer can be evaluated by comparing its buckling resistance with an empty FG cylindrical shell having equal diameter and mass. For obtaining the equivalent thickness h_{eq}, of an empty FG cylindrical shell with a two-layered FG cylindrical shell and equal mass and radius, the following equation can be written:

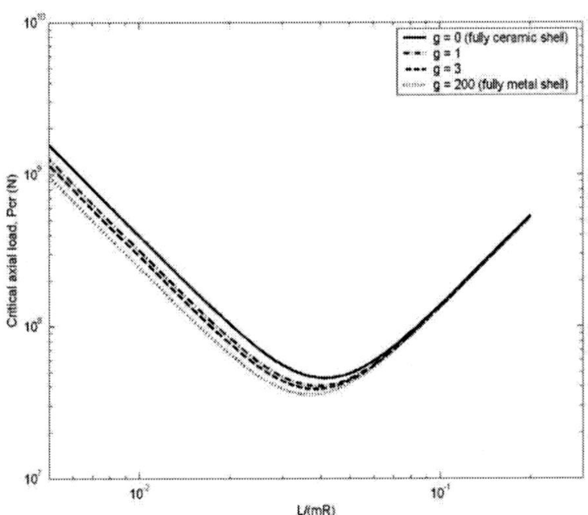

Figure 7. The variation of axial compressive buckling load of two-layered FG cylindrical shell versus L/mR for different values of index factor, g (n=1), (R/t=100), (L/R=2).

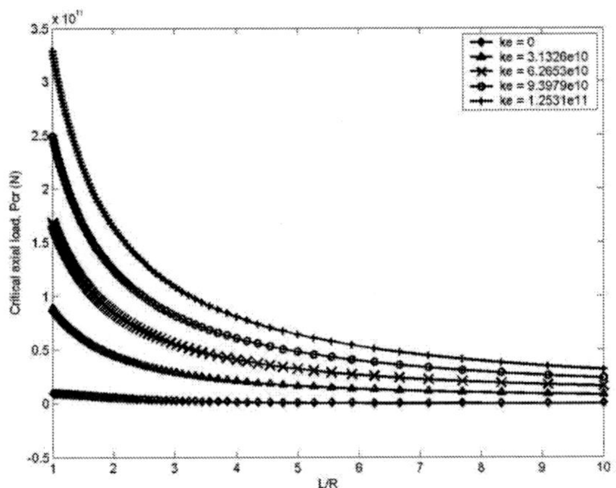

Figure 8. the effect of the elastic layer on the buckling behavior of the FG cylindrical shell, (R/t=100).

$$\rho_{eq} \times 2\pi R h_{eq} = \rho_{eq} \times 2\pi R h + \rho_e \times \pi(R^2 - a^2) \tag{30}$$

where a is the outer radius of the elastic isotropic layer and ρ_{eq} obtained from the following Eq.

$$\rho_{eq}(T,z) = \frac{1}{h}\int_{-h/2}^{h/2} [(\rho_c(T) - \rho_m(T))(z/h + 1/2)^g + \rho_m(T)]dz \tag{31}$$

Therefore

$$h_{eq} = h\left[1 + \frac{t}{2h}\frac{\rho_e}{\rho_{eq}}\left(2 - \frac{t}{R}\right)\right] \tag{32}$$

Figure 9 plots the ratio of elastic buckling load for uniaxial loading of a two-layered FG cylindrical shell to that of a single layered FG cylindrical shell with an equal mass and radius, plotted against the ratio of functionally graded shell radius to thickness for the two-layered FG cylindrical shell. The results are shown for the case in which the inner shell is fully metal, ($g = \infty$). This figure shows that the ratio of the critical loads, $P_{cr}/(P_{cr})_{eq}$, linearly increases as the shell radius to thickness ratio *R/h* increases. Same diagram is shown in the Figure 10 in which the ratio of the nondimensionalized frequency of two-layered FG cylindrical shell to FG cylindrical shell without additional layer is plotted. This figure shows that the ratio of the nondimensionalized frequency, $(\Omega_{\text{with elastic layer}}/\Omega_{\text{without elastic layer}})$, increases as the shell thickness to radius ratio *h/R* increases. It comes to the conclusion that, the presence of additional elastic layers significantly decreases the cost for manufacturing and increases reliability of the products.

CONCLUSION

The free vibration and buckling of a two-layered FG cylindrical shell subjected to combined static and periodic axial forces has been studied in this chapter.

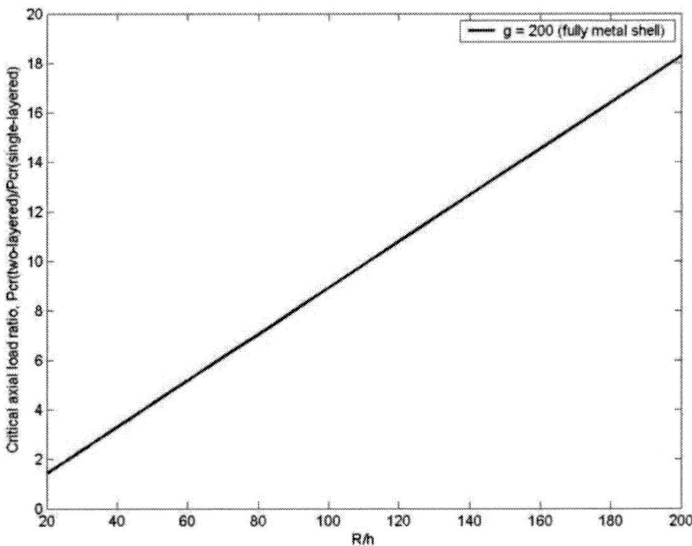

Figure 9. The ratio of elastic buckling load for uniaxial loading of a two-layered FG cylindrical shell to that of a single layered FG cylindrical shell with an equal mass and radius (g=∞).

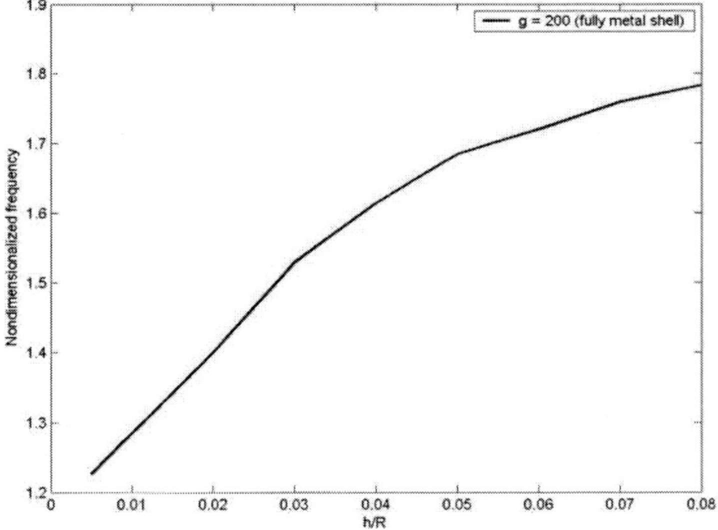

Figure 10. The ratio of nondimensionalized frequency of a two-layered FG cylindrical shell to that of a single layered FG cylindrical shell with an equal mass and radius (g=∞).

The model consists of one isotropic elastic cylindrical shell with functionally graded cylindrical core called two-layered FG cylindrical shell. The new features of the effect of elastic layer on free vibration and buckling of two-layered FG cylindrical shell and some meaningful results in this chapter are helpful for the application and the design of nuclear reactors, space planes and chemical plants, in which FG cylindrical layer act as basic element.

In this research, material properties of functionally graded cylindrical shell are considered as temperature dependent and graded in the thickness direction according to a power-law distribution in terms of the volume fractions of the constituents. A comparison is made to show the free vibration behavior of the structure through results obtained by using two different methods such as the first-order shear deformation theory considering the transverse shear strains and the rotary inertias and the classical shell theory. It can be inferred from the results that the fundamental frequency of two-layered FG cylindrical shell subjected to combined static and periodic axial forces is dependent on the material composition, the deformation mode, and geometry parameters of the shell. Also it can be observed that the transverse shear and rotary inertias have considerable effect on the fundamental frequency of the structure.

The elastic buckling of the two-layered FG cylindrical shell also has been analyzed, and its buckling resistance was compared with that of a FG hollow cylindrical shell. Results show that the critical axial load decreases with increasing FGM index factor and the results are validated using data obtained via FEM. In addition, the application of an elastic layer increases elastic stability of the structure.

REFERENCES

[1] Ebrahimi F, Rastgoo A. Free vibration analysis of smart annular FGM plates integrated with piezoelectric layers. *Smart Mater. Struct.* 2008;17:015044.

[2] Ebrahimi F, Rastgoo A. An analytical study on the free vibration of smart circular thin FGM plate based on classical plate theory. *Thin-Walled Structure*s 2008;46:1402-1408.

[3] Shen HS, Chen TY. A boundary layer theory for the buckling of thin cylindrical shells under axial compression. In: Chien WZ, Fu ZZ, editors. *Advances in Applied Mathematics and Mechanics in China*. Vol. 2. Beijing, China: *International Academic Publishers*, 1990;155–72.

[4] Shen HS. Post-buckling analysis of imperfect stiffened laminated cylindrical shells under combined external pressure and axial compression. *Computers and Structures* 1997;63:335–48.

[5] Ghorbanpour Arani A, Golabi S, Loghman A, Daneshi H, Investigating Elastic Stability of Cylindrical Shell with an Elastic Core Under Axial Compression by Energy Method. *Journal of Mechanical Science and Technology* 2007;21:983-996.

[6] Hutchinson JW, He MY. Buckling of Cylindrical Sandwich Shells with Metal Foam Cores. 2000;37: 6777-6794.

[7] Agarwal BL, Sobel LH. Weight Comparisons of Optimized Stiffened, Unstiffened, and Sandwich Cylindrical Shells. *AIAA J.* 1977;14:1000-1008.

[8] Xiang LIU, CHEN Ji-ming, ZHANG Fu, Zeng-yu XU, Chang-chun GE, Jiang-tao LI. High Heat Flux Testing of B4C/Cu and SiC/Cu Functionally Graded Materials Simulated by Laser and Electron Beam, Plasma. *Science and Technology* 2002; 4:No.1.

[9] Baluc N, Abe K, Boutard JL, Chernov VM, Diegele E, Jitsukawa S, et al. Status of RandD activities on materials for fusion power reactors. *Nucl. Fusion* 2007;47:S696–S717.

[10] Qian Wu, Liu GR, Chun LU, Lam KY. Active vibration control of composite laminated cylindrical shells via surface-bonded magnetostrictive layers. *Smart Mater. Struct.* 2003;12: 889–897.

[11] Alibeigloo A. Static analysis of a functionally graded cylindrical shell with piezoelectric layers as sensor and actuator. *Smart Mater. Struct.* 2009;18:065004-12.

[12] Gough GJ, Elam CF, de Bruyne NA. The Stabilization of a Thin Sheet by a Continuous Supporting Medium. *The Journal of Royal Aeronautical Society* 1940;44: 12-43.

[13] Jafari AA, Khalili SMR, Azarafza R. Transient dynamic response of composite circular cylindrical shells under radial impulse load and axial compressive loads. Thin-Walled Structures 2005;43:1763-86.

[14] Reddy JN. Mechanics of Laminated Composite Plates and Shells. Second Edition, CRC Press, New York;2004.

[15] Radwan HR, Genin J. Dynamic instability in cylindrical shells. *Journal of Sound and Vibration* 1978; 56:373-82.

Chapter 4

ELASTIC STABILITY ANALYSIS OF A TWO-LAYERED FUNCTIONALLY GRADED CYLINDRICAL SHELL UNDER AXIAL COMPRESSION

ABSTRACT

This chapter investigates the elastic axisymmetric buckling of a thin, simply supported functionally graded (FG) cylindrical shell embedded with an elastic layer under axial compression. The analysis is based on energy method and simplified nonlinear strain-displacement relations for axial compression. Material properties of functionally graded cylindrical shell are considered graded in the thickness direction according to a power-law distribution in terms of the volume fractions of the constituents. Using minimum potential energy together with Euler equations, equilibrium equations are obtained. Consequently, stability equation of functionally graded cylindrical shell with an elastic layer is acquired by means of minimum potential energy theory and Trefftz criteria. Another analysis is made using the equivalent properties of FG material. Numerical results for stainless steel-ceramic cylindrical shell and aluminum layer are obtained and critical load curves are analyzed for a cylindrical shell with an elastic layer. A comparison is made to the results in the literature. The results show that the elastic stability of functionally graded cylindrical shell with an elastic layer is dependent on the material composition and FGM index factor, and the shell geometry parameters and it is concluded that the application of an elastic layer increases elastic stability and significantly reduces the weight of cylindrical shells.

1. INTRODUCTION

As stated before functionally graded materials (FGMs) have received considerable attention in many engineering applications since they were first reported in 1984. FGMs are composite materials, microscopically inhomogeneous, in which the mechanical properties vary smoothly and continuously from one surface to the other. This is achieved by gradually varying the volume fraction of the constituent materials. Stein [1] was the first to recognize the importance of nonlinear prebuckling deformations on the buckling load of perfect cylindrical shells. It has been shown by Shen and Chen [2,3] that in shell buckling, there is a boundary layer phenomenon where prebuckling and buckling displacement vary rapidly. They suggested a boundary layer theory of shell buckling, which includes the effects of nonlinear prebuckling deformations, large deflections in the postbuckling range, and initial geometric imperfections of the shell. Based on this theory, the postbuckling analyses for perfect and imperfect, unstiffened and stiffened, isotropic and multilayered cylindrical shells under various loading cases have been performed by Shen and Chen [4], Shen et al. [5], and Shen [6–10]. It should be noted that in some of above studies the material properties are assumed to be independent of the temperature.

Karam et al. [11] analyzed the elastic buckling of a thin cylindrical shell supported by an elastic core. The results they reported show that this structural configuration achieves significant weight saving compared with a hollow cylinder. Agrawal et al. [12] determined the weight advantage of honeycomb sandwich cylindrical shells under axial compression compared with axially stiffened cylinders. By comparing weight optimized shells, they showed that honeycomb sandwiches offer a substantial weight advantage over a significant load range. Hutchinson et al. [13] studied buckling of cylindrical shells with metal foam layers subjected to axial compression. The results reported suggest that for sandwich shells and curved sandwich panels there is a practically significant loading range for which optimized sandwich constructions with metal foam cores, have a distinct weight advantage over metal stringer constructions. However, these results also indicate that the imperfection sensitivity is essentially as severe as it is for the monocoque cylinder.

This chapter studies the buckling of a two-layered functionally graded cylindrical shell subjected to axial forces. The assumed model in this study consists of one isotropic elastic cylindrical shell with functionally graded cylindrical core called two-layered FGCL. The elastic buckling of the two-layered FGCS has been analyzed, and its buckling resistance was compared with that of a FGM hollow cylindrical shell. Material properties of functionally graded cylindrical shells are considered as temperature dependent and graded in the thickness direction according to a power-law distribution in terms of the volume fractions of the constituents. It is found that the effect of elastic layer on buckling of two-layered FGCS subjected to axial force is not neglected in some cases. The new features of the effect of elastic layer on the buckling of two-layered FGCS and some meaningful results in this chapter are helpful for the application and the design of nuclear reactors, space planes and chemical plants, in which FG cylindrical layer act as basic element.

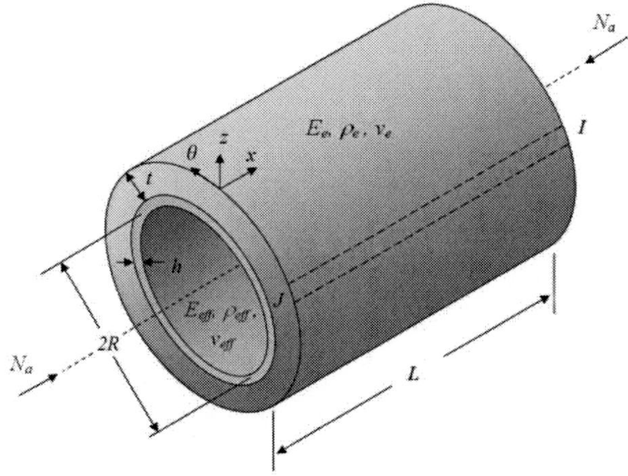

Figure 11. Geometry of a two-layered functionally graded cylindrical shell with loading configuration.

2. Formulation of the Problem

2.1. Analysis

An Functionally Graded (FG), closed-ended, perfectly circular cylindrical shell embedded with an isotropic elastic layer with simply supported end conditions has been considered in this research. The functionally graded layer has a uniform thickness h, radius R, length L, density $\rho_{eff}(T,z)$, Young's modulus $E_{eff}(T,z)$ and Poisson ratio $v_{eff}(T,z)$. The properties are assumed to be graded in the thickness direction according to a power-law distribution in terms of the volume fractions of the constituents. The isotropic elastic layer's properties are: density ρe, Young's modulus Ee, Poisson ratio υe, and a uniform thickness t as shown in Figure 1. Coordinates x, θ, and z axes are defined with corresponding displacements u, v and w.

The middle-surface kinematic relations on which the Donnel equations are based are [14]

$$\varepsilon_x^0 = u_{,x} + \frac{1}{2}\beta_x^2 , \quad \varepsilon_y^0 = \frac{v_{,\theta} + w}{a} + \frac{1}{2}\beta_y^2 , \quad \gamma_{x\theta}^0 = \left(\frac{u_{,\theta}}{a} + v_{,x}\right) + \beta_x \beta_\theta \tag{1}$$

$$k_x = \beta_{x,x} , \quad k_\theta = \frac{\beta_{\theta,\theta}}{R} , \quad k_{x\theta} = \left(\frac{\beta_{x,\theta}}{R} + \beta_{\theta,x}\right)$$

in which $\beta_x = -w_{,x}$ and $\beta_\theta = -w_{,\theta}/R$. The stress–strain relations are given as

$$\begin{aligned} \sigma_x &= K(z)[\varepsilon_x + v(z)\varepsilon_\theta] \\ \sigma_\theta &= K(z)[\varepsilon_\theta + v(z)\varepsilon_x] \\ \tau_{x\theta} &= K(z)[1 - v(z)]\gamma_{x\theta} \end{aligned} \tag{2}$$

where $\mathbf{K(z) = E(z)/(1 - v(z)^2)}$. For thin cylindrical shells $(h/R) \ll 1$, the approximate forms of internal force and moment related to σ_{ij} through the shell thickness based on the first-order shell theory are

$$\{(N_x, N_\theta, N_{x\theta}), (M_x, M_\theta, M_{x\theta})\} = \int_{-h/2}^{h/2} (\sigma_x, \sigma_\theta, \tau_{x\theta}) \tag{3}$$

Considering Eqs. (1) and (2) in Eq. (3) yields

$$\begin{Bmatrix} N \\ M \end{Bmatrix} = \begin{bmatrix} A & D \\ D & B \end{bmatrix} \begin{Bmatrix} \varepsilon \\ k \end{Bmatrix} \tag{4}$$

in which $\{N\ M\}^T = \{N_x\ N_\theta\ N_{x\theta}\ M_x\ M_\theta\ M_{x\theta}\}^T$,

$\{\varepsilon\ k\}^T = \{\varepsilon_x\ \varepsilon_\theta\ \gamma_{x\theta}\ k_x\ k_\theta\ 2k_{x\theta}\}^T$ and the Poison's ratio is assumed to be constant, $v(z) = v$, through the shell thickness because of thin cylindrical shell, $(h/R) \ll 1$. Therefore

$$A = I_0.q\ ,\ \ D = I_1.q\ ,\ \ B = I_2.q \tag{5}$$

where

$$q = \begin{bmatrix} 1 & v & 0 \\ v & 1 & 0 \\ 0 & 0 & \dfrac{1-v}{2} \end{bmatrix} \tag{6}$$

and

$$I_i = \frac{1}{1-v^2}\int_{-h/2}^{h/2} z^i E(z)dz \tag{7}$$

For isotropic cylindrical shells, we have $I_1 = 0$, so $D = 0$. Note that *A*, *B* and *D* are symmetric matrixes and $A_{ij} = A_{ji}, B_{ij} = B_{ji}, D_{ij} = D_{ji}$.

2.2. Functionally Graded Materials (FGM)

Several available analytical and computational models have discussed the issue of finding suitable functions for material properties, and there are several criteria for selecting them. They are desired to be continuous, simple and should have the ability to exhibit curvature, both 'concave upward' and 'concave downward'. In this study the simple power law, which has all the desired properties, is used. Nowadays not only FGM can easily be produced but also one can control even the variation of the FG constituents in a specific way. For example in an FG material made of a ceramic and metal mixture, we have

$$V_m + V_c = 1 \tag{8}$$

in which V_c and V_m are the volume fraction of the ceramic and metallic part, respectively. Based on the power law distribution, the variation of V_c versus thickness coordinate (z) with its origin placed at the middle of the thickness can be expressed as

$$V_c(z) = (z/h + 1/2)^g \ , \ g \geq 0 \tag{9}$$

in which h is the FG core layer thickness and g is the FGM index factor. Note that the variation of both the ceramic and metal constituents is linear when g = 1. Moreover, for the value of g = 0, a fully ceramic plate is intended. All other mechanical, physical and thermal properties of FGM media follow the same distribution as for *Vc*. We assume that the inhomogeneous material properties, such as the modulus of elasticity *E,* the density ρ, thermal expansion coefficient α, and thermal diffusivity k, change within the thickness direction z based on Voigt's rule over the whole range of the volume fraction as follows, while Poisson's ratio υ is assumed to be constant in the thickness direction as

$$\begin{aligned} E(z) &= (E_c - E_m)V_c(z) + E_m \\ \rho(z) &= (\rho_c - \rho_m)V_c(z) + \rho_m \\ \alpha(z) &= (\alpha_c - \alpha_m)V_c(z) + \alpha_m \\ k(z) &= (k_c - k_m)V_c(z) + k_m \\ v(z) &= v \end{aligned} \tag{10}$$

where the subscripts m and c refer to the metal and ceramic constituents, respectively. After substituting *Vc* from Eq. (2) into Eq. (3), material properties of the FGM layer can be determined in the power law form which are the same as those proposed by Reddy and Praveen, i.e.

$$\begin{aligned} E(z) &= (E_c - E_m)(z/h_2 + 1/2)^g + E_m \\ \rho(z) &= (\rho_c - \rho_m)(z/h_2 + 1/2)^g + \rho_m \\ &etc. \end{aligned} \tag{11}$$

Contrary to most published papers on FG structures, in this chapter we choose a shell with the outer part as being metallic and the inner part as ceramic. This makes us exchange the coordinate direction in Figure 1, so that z-axis passes the center of the shell.

2.3. Energy Formulation

The total potential energy of the FGCS with an isotropic elastic outer layer under axial compressive force is the sum of membrane strain energy U_m, the bending strain energy U_b, the membrane-bending energy U_s, the strain energy stored in the elastic layer U_e, and the potential energy Π of the applied compressive force expressed as

$$V = U_m + U_b + U_s + U_e + \Pi \tag{12}$$

where:

$$U_m = \frac{R}{2}\iint \left[A_{11}.\varepsilon_x^{o\,2} + A_{22}.\varepsilon_\theta^{o\,2} + 2A_{12}.\varepsilon_x\varepsilon_\theta + A_{33}.\gamma_{x\theta}^{o\,2}\right]dx\,d\theta$$
$$U_b = \frac{R}{2}\iint [B_{11}.k_x^2 + B_{22}.k_\theta^2 + 2B_{12}.k_x k_\theta + B_{33}.k_{x\theta}^2]dx\,d\theta$$
$$U_s = R\iint [D_{11}.\varepsilon_x k_x + D_{12}.\varepsilon_x k_\theta + D_{12}.\varepsilon_\theta k_x + D_{22}.\varepsilon_\theta k_\theta + D_{33}.\varepsilon_{x\theta} k_{x\theta}]dx\,d\theta$$
$$U_e = \frac{k_e}{2}\iint w^2 R\,dx\,d\theta$$
$$\Pi = -\iint (P_x u)R\,dx\,d\theta \tag{13}$$

In which *ke* is the spring constant for the compliant elastic layer. For zero spring stiffness and isotropic shell this reduces to the result for a hollow cylindrical shell. Assuming that the shell thickness is much smaller than its radius, the spring constant k_e can be found from the result of stress in the z direction of a flat strip element (*IJ* in Figure 1) with an elastic foundation subjected to a sinusoidal displacement in the z direction [15]:

$$\sigma_z = -\frac{2\pi E_e}{(3-\upsilon_e)(1+\upsilon_e)} \times \frac{m}{L} w \tag{14}$$

where σ_z, m and L are stress in the z direction, longitudinal wave number and shell length, respectively. For a cylindrical shell with an outer elastic layer under external pressure q, we have

$$q = \sigma_z = -k_e w \tag{15}$$

where w is the radial displacement of the elastic layer. Equations (6) and (7) lead to

$$k_e = -\frac{2E_e}{(3-\upsilon_e)(1+\upsilon_e)} \times \frac{1}{\lambda} \tag{16}$$

where λ is buckling wavelength parameter

$$\lambda = \frac{L}{m\pi} \tag{17}$$

Minimizing the potential energy function by means of the Euler equations, and using Eqs. (4), (12) and (13), the equilibrium equations for thin cylindrical shell with a elastic layer (two-layered FGCS) are obtained as

$$\begin{aligned}
&RN_{x,x} + N_{x\theta,\theta} = -RP_x \\
&RN_{x\theta,x} + N_{\theta,\theta} = 0 \\
&M_{x,xx} + \frac{2}{R}M_{x\theta,x\theta} + \frac{1}{R^2}M_{\theta,\theta\theta} - \frac{1}{R}N_\theta - \left(N_x w_{,xx} + \frac{2}{R}N_{x\theta}(w_{,x\theta}) + \frac{1}{R}N_\theta w_{,\theta\theta}\right) + k_e w \\
&= 0
\end{aligned} \tag{18}$$

2.4. Stability Equation

If V is the total potential energy of the shell, its variation in equilibrium state using the Taylor series leads to

$$\Delta V = \delta V + \frac{1}{2!}\delta^2 V + \frac{1}{3!}\delta^3 V + \cdots \tag{19}$$

The condition $\delta^2 V = 0$ shows the stability of the original configuration of the shell in the neighborhood of equilibrium state and is used to derive the stability equations for many buckling problems as discussed by Langhaar [16].

If the state of stable equilibrium of a general cylindrical shell with a layer under axial compressive load P is designated by u_0, v_0 and w_0, the linear displacement portion of the neighboring state are designated by u_1, v_1 and w_1. Similarly the linear portion of the components of forces and moments related to the neighboring state are depicted by subscript *1*. In matrix form we have

$$\begin{Bmatrix} N_0 \\ M_0 \end{Bmatrix} = \begin{bmatrix} A & D \\ D & B \end{bmatrix} \begin{Bmatrix} \varepsilon_0 \\ k_0 \end{Bmatrix} \text{ and } \begin{Bmatrix} N_1 \\ M_1 \end{Bmatrix} = \begin{bmatrix} A & D \\ D & B \end{bmatrix} \begin{Bmatrix} \varepsilon_1 \\ k_1 \end{Bmatrix} \tag{20}$$

in which

$$\begin{aligned} &\{N_1\ M_1\}^T = \{N_{x1}\ N_{\theta 1}\ N_{x\theta 1}\ M_{x1}\ M_{\theta 1}\ M_{x\theta 1}\}^T \\ &\{\varepsilon_1\ k_1\}^T = \{\varepsilon_{x1}\ \varepsilon_{\theta 1}\ \gamma_{x\theta 1}\ k_{x1}\ k_{\theta 1}\ k_{x\theta 1}\}^T \end{aligned} \tag{21}$$

where now the linearized form of the strain-displacement relations are [14]

$$\begin{aligned} &\varepsilon_{x1} = u_{1,x}\ , \quad \varepsilon_{\theta 1} = \frac{v_{1,\theta} + w_1}{R}\ , \quad \gamma_{x\theta 1} = \left(\frac{u_{1,\theta}}{R} + v_{1,x}\right) \\ &k_{x1} = \beta_{x1,x}\ , \quad k_{\theta 1} = \frac{\beta_{\theta 1,\theta}}{R}\ , \quad k_{x\theta 1} = \left(\frac{\beta_{x1,\theta}}{R} + \beta_{\theta 1,x}\right) \end{aligned} \tag{22}$$

Employing $\frac{1}{2}\delta^2 V$, calculated according to above mentioned definitions, into the Euler equations, and neglecting the transverse shear force Q_θ and rotations β_x and β_θ, the Donnel stability equation are obtained

$$\begin{aligned} &RN_{x1,x} + N_{x\theta 1,\theta} = 0 \\ &RN_{x\theta 1,x} + N_{\theta 1,\theta} = 0 \\ &M_{x1,xx} + \frac{2}{R}M_{x\theta 1,x\theta} + \frac{1}{R^2}M_{\theta 1,\theta\theta} - \frac{1}{R}N_{\theta 1} + \left(N_{x0}w_{1,xx} + \frac{2}{R}N_{x\theta 0}w_{1,x\theta} + \frac{1}{R^2}N_{\theta 0}w_{1,\theta\theta}\right) \\ &\quad + k_e w_1 = 0 \end{aligned} \tag{23}$$

3. Buckling of a Two-Layered FGCS Under Axial Compression

By substituting N_1 and M_1 with their strain equivalences from Eq. (22), the Donnel equations for a FGCS embedded with elastic external layer are obtained in terms of displacements. The equations are the stability equations in coupled form with the variables u_1, v, w_1 and may be partially uncoupled and written in the following form:

$$\nabla^4 u_1 = -\frac{v}{R} w_{1,xxx} + \frac{1}{R^3} w_{1,x\theta\theta} - \left(\frac{I_1}{I_0}\right) \nabla^4 w_{1,x}$$
$$\nabla^4 v_1 = -\frac{2+v}{R^2} w_{1,xx\theta} - \frac{1}{R^4} w_{1,\theta\theta\theta} - \left(\frac{I_1}{I_0}\right) \nabla^4 w_{1,\theta}$$
$$I_2 \nabla^8 w_1 = \nabla^4 \left(R N_{\theta 1} - \left[R^2 N_{x0} w_{1,xx} + 2R N_{x\theta 0} w_{1,x\theta} + N_{\theta 0} w_{1,\theta\theta}\right] - R^2 k_e w_1\right)$$
$$- I_1 \nabla^4 \left(R^2 u_{1,xxx} + u_{1,x\theta\theta} + \frac{v}{R}(v_{1,xx\theta} + w_{1,xx}) + R(1-v) v_{1,xx\theta} + \frac{1}{R}(v_{1,\theta\theta\theta} + w_{1,\theta\theta})\right)$$
(24)

In above equation all quadratic and higher order terms in *u*, *v*, *w* are omitted in right-hand-side. A simply supported two-layered FGCS subjected to a uniformly distributed axial compressive load *P* is considered. The initial deformation is axisymmetric, and the axial compressive buckling load of shell with an elastic layer P_{cr}, is the lowest load at which equilibrium in the axisymmetric form ceases to be stable. From a membrane analysis of the unbuckled cylindrical form [14], the prebuckling forces are

$$N_{x0} = -\frac{P}{2\pi R}, \quad N_{x\theta 0} = N_{\theta 0} = 0 \tag{25}$$

Substituting prebuckling forces into the third Eq. (24) yields:

$$I_2 \nabla^8 w_1 = \nabla^4 \left(R - R^2 \left(\frac{-P}{2\pi R}\right) w_{1,xx} - R^2 k_e w_1\right)$$
$$- I_1 \nabla^4 \left(R^2 u_{1,xxx} + u_{1,x\theta\theta} + \frac{v}{R}(v_{1,xx\theta} + w_{1,xx}) + R(1-v) v_{1,xx\theta} + \frac{1}{R}(v_{1,\theta\theta\theta} + w_{1,\theta\theta})\right)$$
(26)

where $\nabla^8 w_1 = \nabla^4(\nabla^4 w_1)$. From the edge conditions

$$w_1 = w_{1,xx} = 0 \tag{27}$$

Assume a solution as:

$$w_1 = C_1 \sin \bar{m}x \sin \bar{n}\theta \tag{28}$$

in which $\bar{m} = m\pi R/L$, and $\bar{n} = n/R$. From first and second Eq. (24)

$$\nabla^4 u_1 = K_1\, C_1 \cos \bar{m}x \sin \bar{n}\theta$$

and

$$\nabla^4 v_1 = K_2\, C_1 \sin \bar{m}x \cos \bar{n}\theta \tag{29}$$

where

$$K_1 = \bar{m}\left[\frac{\nu}{R}\bar{m}^2 - \frac{1}{R}\bar{n}^2 - \frac{I_1}{I_0}(\bar{m}^4 + 2\bar{m}^2\bar{n}^2 + \bar{n}^4)\right] \tag{30}$$

and

$$K_2 = \bar{n}\left[\frac{2+\nu}{R}\bar{m}^2 + \frac{1}{R}\bar{n}^2 - \frac{I_1}{I_0}R(\bar{m}^4 + 2\bar{m}^2\bar{n}^2 + \bar{n}^4)\right] \tag{31}$$

and then

$$u_1 = C_2 \cos \bar{m}x \sin \bar{n}\theta \text{ and } v_1 = C_3 \sin \bar{m}x \cos \bar{n}\theta \tag{32}$$

where

$$C_2 = C_1\, K_1\, (\bar{m}^4 + 2\bar{m}^2\bar{n}^2 + \bar{n}^4)^{-1} \text{ and}$$
$$C_3 = C_1\, K_2 (\bar{m}^4 + 2\bar{m}^2\bar{n}^2 + \bar{n}^4)^{-1} \tag{33}$$

Inserting u_1 and v_1 into third Eq. (26) gives

$$I_2\nabla^8 w_1 + I_0\frac{(1-\nu^2)}{R^2}w_{1,xxxx} + \nabla^4\left(\frac{P}{2\pi R}w_{1,xx} + (k_e - I_1C_4)w_1\right) = 0 \tag{34}$$

in which

$$C_4 = \left[\frac{(K_2\bar{n} - K_1\bar{m})}{(\bar{m}^2 + \bar{n}^2)^2} - \frac{v}{R}\right](\bar{m}^2 - \bar{n}^2) \tag{35}$$

Inserting Eqs. (28) and (32) into Eq. (34) gives

$$\frac{P}{2\pi R} = \frac{I_2(\bar{m}^2 + \bar{n}^2)^2 + \frac{I_0}{R^2}(1 - v^2)\frac{\bar{m}^4}{(\bar{m}^2 + \bar{n}^2)^2} + k_e + I_1\left[\frac{(K_2\bar{n} - K_1\bar{m})}{(\bar{m}^2 + \bar{n}^2)^2} - \frac{v}{R}\right](\bar{m}^2 - \bar{n}^2)}{\bar{m}^2} \tag{36}$$

In Eq. (36), the minimum axial load is the critical axial load P_{cr}, which can be obtained by specifying values for *m* and *n*. Solving the equation for the two-layered isotropic case the result is

$$\frac{P}{2\pi R} = \frac{D(\bar{m}^2 + \bar{n}^2)^2 + \frac{1 - v^2}{R^2}C\frac{\bar{m}^4}{(\bar{m}^2 + \bar{n}^2)^2} + k_e}{\bar{m}^2} \tag{37}$$

4. Formulation with Use of FG Equivalent Properties

Using the Eq. (8) gives rise to the effective properties of functionally graded materials. The corresponding effective material properties for the FGCS that is ceramic rich at the inner surface and metal rich at the outer surface are expressed as

$$F_{eff}(T, z) = F_c(T)V_c(z) + F_m(T)(1 - V_c(z)) \tag{38}$$

Here subscript *c* refers to the ceramic, subscript *m* refers to the metal and F_{eff} is the effective material property of the FGCS, including the effective elastic modulus, effective mass density and effective Poisson's ratio. F_c and F_m are the temperature dependent properties of the ceramic and metal, respectively. On the other hand, an FGCS that is ceramic rich at the inner surface and metal rich at the outer surface is defined as Type B, (in contrary to Type A, which is metal rich at the inner surface and ceramic rich at the outer surface) whose effective material properties are given by

$$E_{eff}(T,z) = (E_c(T) - E_m(T))(z/h + 1/2)^g + E_m(T)$$
$$\rho_{eff}(T,z) = (\rho_c(T) - \rho_m(T))(z/h + 1/2)^g + \rho_m(T)$$
$$\nu_{eff}(T,z) = (\nu_c(T) - \nu_m(T))(z/h + 1/2)^g + \nu_m(T) \tag{39}$$

The equivalent properties are described as

$$F_{eq}(T,z) = \frac{1}{h}\int_{-h/2}^{h/2} F_{eff}(T,z)dz = \frac{1}{h}\int_{-h/2}^{h/2} [(F_c(T) - F_m(T))(z/h + 1/2)^g + F_m(T)]dz \tag{40}$$

The initial deformation is axisymmetric, and the axial compressive buckling load of a FGCS with an elastic layer P_{cr}, is the lowest load at which equilibrium in the axisymmetric form ceases to be stable. Following a procedure similar to that of section 3, the uncoupled stability equations are obtained. From Eq. (23) we have

$$\nabla^4 u_1 = -\frac{\nu_{eq}}{R} w_{1,xxx} + \frac{1}{R^3} w_{1,x\theta\theta}$$
$$\nabla^4 v_1 = -\frac{2+\nu_{eq}}{R^2} w_{1,xx\theta} - \frac{1}{R^4} w_{1,\theta\theta\theta}$$
$$\mathbb{D}\nabla^8 w_1 + \left(\frac{1-v_{eq}^2}{R^2}\right)\mathbb{C}.\, w_{1,xxxx} - \nabla^4\left[N_{x0} w_{1,xx} + \frac{2}{R} N_{x\theta 0} w_{1,x\theta} + \frac{1}{R^2} N_{\theta 0} w_{1,\theta\theta} - k_e w_1\right] = 0 \tag{41}$$

in which $\mathbb{C} = E_{eq}h/(1-v_{eq}^2)$ and $\mathbb{D} = E_{eq}h^3/12(1-v_{eq}^2)$. Substituting prebuckling forces from Eq. (25) into the third Eq. (41) yields:

$$\mathbb{D}\nabla^8 w_1 + \left(\frac{1-v_{eq}^2}{R^2}\right)\mathbb{C}.\, w_{1,xxxx} + \nabla^4\left[\frac{P_x w_{1,xx}}{2\pi R} + k_e w_1\right] = 0 \tag{42}$$

Equation (42) has constant coefficients and from the edge conditions:

$$w_1 = w_{1,xx} = 0 \tag{43}$$

Assume a solution as:

$$w_1 = C_1 \sin \bar{m}x \sin \bar{n}\theta \tag{44}$$

where $\bar{m} = m\pi R/L$ and $\bar{n} = n/R$; *m* and *n* are the axial half-wave number and the circumferential full-wave number , respectively, and C_1 is the amplitude of the displacement function. Substituting the solution w_1 from Eq. (44) into Eq. (42) results in:

$$\mathbb{D}(\bar{m}^2+\bar{n}^2)^4 w_1 + \frac{1-v_{eq}^2}{R^2}\mathbb{C}\bar{m}^4 w_1 - \frac{P}{2\pi R}\bar{m}^2(\bar{m}^2+\bar{n}^2)^2 w_1 + k_e(\bar{m}^2+\bar{n}^2)^2 w_1 = 0 \tag{45}$$

Dividing Eq. (45) by $(\bar{m}^2+\bar{n}^2)^2 \times w$ and rearranging it gives:

$$\frac{P}{2\pi R} = \frac{\mathbb{D}(\bar{m}^2+\bar{n}^2)^2 + \dfrac{1-v_{eq}^2}{R^2}\mathbb{C}\dfrac{\bar{m}^4}{(\bar{m}^2+\bar{n}^2)^2} + k_e}{\bar{m}^2} \tag{46}$$

5. Influence of Elastic Layer

The performance of a thin walled FGCS with a compliant elastic layer can be evaluated by comparing its buckling resistance with an empty FGCS having equal diameter and mass. For calculating the equivalent thickness h_{eq}, of an empty FGCS with a two-layered FGCS and equal mass and radius, the following equation can be written:

$$\rho_{eq} \times 2\pi R h_{eq} = \rho_{eq} \times 2\pi R h + \rho_e \times \pi(R^2 - a^2) \tag{47}$$

where *a* is the outer radius of the elastic isotropic layer and ρ_{eq} obtained from Eq. (40).

Therefore

$$h_{eq} = h\left[1 + \frac{t}{2h}\frac{\rho_e}{\rho_{eq}}\left(2 - \frac{t}{R}\right)\right] \tag{48}$$

9. RESULTS AND DISCUSSION

The ceramic material used in this case is silicon nitride and the metal material used is ASTM-A723 steel. The elastic layer is assumed to be from aluminum 7070-T7651. The material properties are tabulated in Table 1.

Table 1. Material properties [26]

Material	Density	Module of Elasticity	Poisson's ratio	Constants in Eq. (49)
Silicon Nitride	$\rho_c = 2370\ kg/m^3$	$E_{0c} = 348.42\ GPa$	$\nu_c = 0.24$	$c_{1c} = 3.07 \times 10^{-4}$ $c_{2c} = 2.16 \times 10^{-7}$ $c_{3c} = 8.946 \times 10^{-11}$
ASTM-A723	$\rho_m = 8900\ kg/m^3$	$E_{0m} = 197\ GPa$	$\nu_m = 0.31$	$c_{1m} = 2.794 \times 10^{-4}$ $c_{2m} = -3.998 \times 10^{-9}$ $c_{3m} = 0$
Zirconia	$\rho_c = 5680\ kg/m^3$	$E_{0c} = 244.27\ GPa$	$\nu_c = 0.24$	$c_{1c} = -1.3107 \times 10^{-3}$ $c_{2c} = 1.2139 \times 10^{-6}$ $c_{3c} = -3.6814 \times 10^{-10}$
Ti-6Al-4V	$\rho_c = 4430\ kg/m^3$	$E_{0m} = 122.56\ GPa$	$\nu_m = 0.33$	$c_{1m} = -4.58635 \times 10^{-4}$ $c_{2m} = 0$ $c_{3m} = 0$
Al 7070-T7651	$\rho_e = 2770\ kg/m^3$	$E_e = 70.0\ GPa$	$\nu_e = 0.3$	

In this study the FGCS has a uniform thickness of h=*5mm*, the radius range of $0.1m \le R \le 1m$ and length $L = 1m$ with an elastic layer with uniform thickness of $t = 5h$. The FGCS is considered to be simply supported at both ends and T is the temperature in Kelvin in the following relations [17].

$$\begin{aligned} E_c &= E_{0c}\left(1 - c_{1c}T + c_{2c}T^2 + c_{3c}T^3\right) \\ E_m &= E_{0m}\left(1 - c_{1m}T + c_{2m}T^2 + c_{3m}T^3\right) \end{aligned} \tag{49}$$

Considering the above specifications and equation (36), the critical force per unit circumferential length of the two-layered FGCS, N_{cr}, versus aspect ratio *R/h* is shown in Figure 2. Results are illustrated for *g=0, 1, 3,200*. It is obvious that as the shell radius increases, the buckling load decreases. The variation of the critical load is more pronounced in smaller amount of aspect ratio.

Besides, one can see the effect of material constituents of FGCS on the critical force. The stability of cylinder increases when the ceramic content increases in FGM composition.

Similarly, the effect of the shell length on the critical axial force of the two-layered FGCS,P_{cr}, is shown in Figure 3, for *g=0, 1, 3,200*. It is obvious that as the shell length increases, the buckling load decreases.

The effect of FGM index factor, g, on the critical load can be better seen in Figure 4. The critical axial force decreases as the FGM index factor increases. The dashed line in this figure shows the critical load obtained by considering FGM effective properties (section 4). It can be inferred from the diagram that the difference between results obtained from Eq. (36) and Eq. (46) is less than 10.0%.

Figure 5 represents the variation of axial compressive buckling load of FGCS with an elastic layer versus L/mR for different values of index factor, g. The horizontal axis shows the value of L/mR and the vertical axis representsN_{cr}. The exact relationship between N_{cr} and L/mR is obtained is ploted for *n=1*. For each value of g, the critical load is minimum. Thus the critical load is the minimum point of the curve in Fig 5. It is also inferred that the minimum critical load is obtained for $g \to \infty$.

Figure 6 represents the effect of the elastic layer on the buckling behavior of the two-layered FGCS. As one can see from Figure 6, the presence of an elastic layer significantly increases the buckling resistance in axial compression with respect to a hollow FGCS. Therefore, the application of an elastic layer increases elastic stability.

Figure 7 plots the ratio of elastic buckling load for uniaxial loading of a two-layered FGCS to that of a single layered FGCS with an equal mass and radius, plotted against the ratio of functionally graded shell radius to thickness for the two-layered FGCS. The results are shown in three cases: the case that the inner shell is fully metal, ($g = \infty$), the case that the inner shell is graded linearly, ($g = 1$), and the case that the inner shell is fully ceramic, ($g = 0$). This figure shows that the ratio of the critical loads, $P_{cr}/(P_{cr})_{eq}$, linearly increases as the shell radius to thickness ratio *R/h* increases. It is concluded that the presence of an elastic layer significantly increases the buckling resistance in axial compression with respect to a hollow FGCS of equal mass and radius.

To validate the equation (36), a comparison is made to the results obtained by Huang and Han [18]. Figure 8 shows the relation curves of critical axial stress, σ_{cr} versus FGM index factor *g*. The structure is considered to be one layer FGCS and the dimensional parameters of the shells are given in the figure as well. The material properties for metal (Ti-6Al-4V) and ceramic (Zirconia) are listed in Table 1. *g*-Axis is in the logarithmic form. The results shown in Figure 8 are in reasonable conformance with the Hung and Han's results shown in Figure 9. As shown, σ_{cr} decreases with the increase of *g*. The prime reason for the fall of σ_{cr} is that a higher value of *g* corresponds to a metal-richer shell, which usually has less stiffness than a ceramic-richer one.

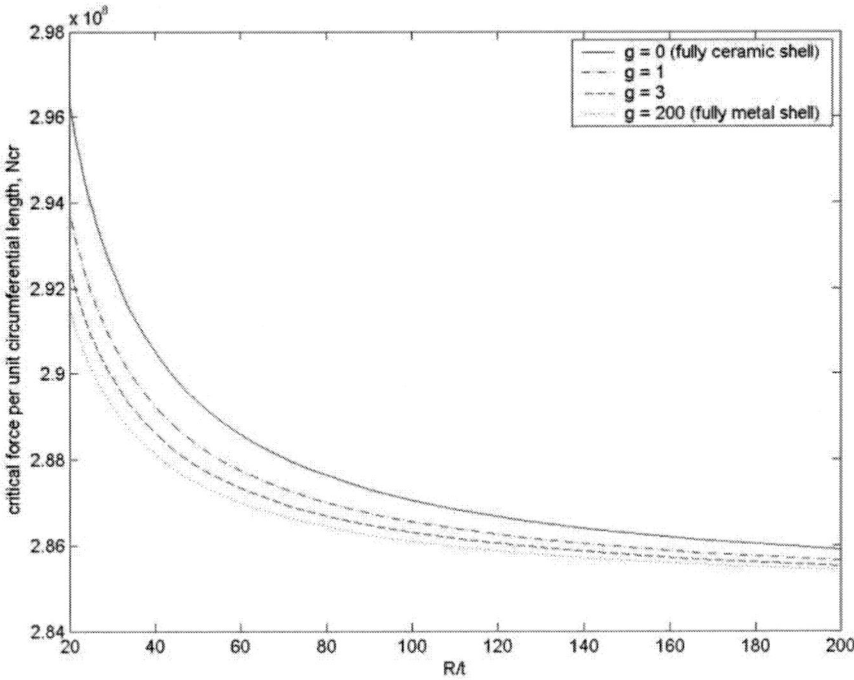

Figure 12. The critical force per unit circumferential length of the FGM cylindrical shell with an elastic layer, N_{cr}, versus aspect ratio R/h, (g=0, 1, 3,200).

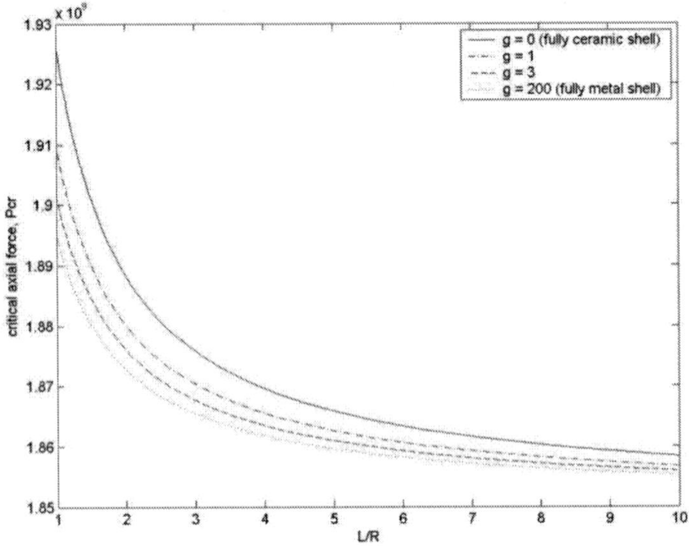

Figure 13. The effect of the shell length on the critical axial force of the FGM cylindrical shell with an elastic layer, P_{cr}.

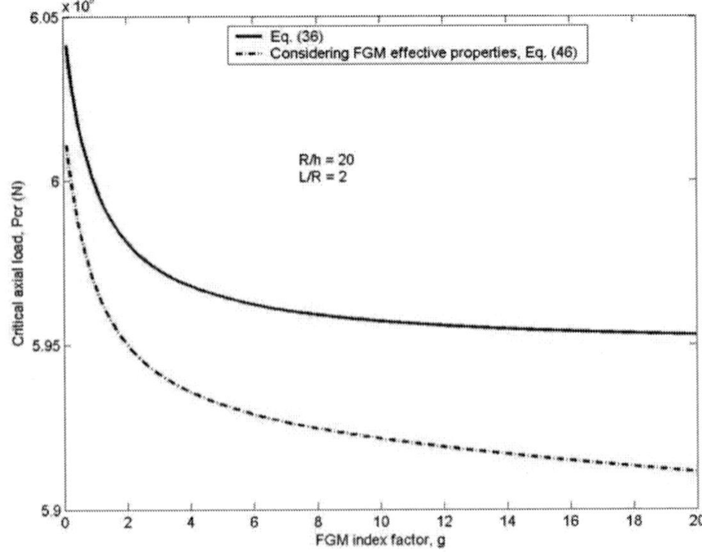

Figure 14. The effect of FGM index factor, g, on the critical load.

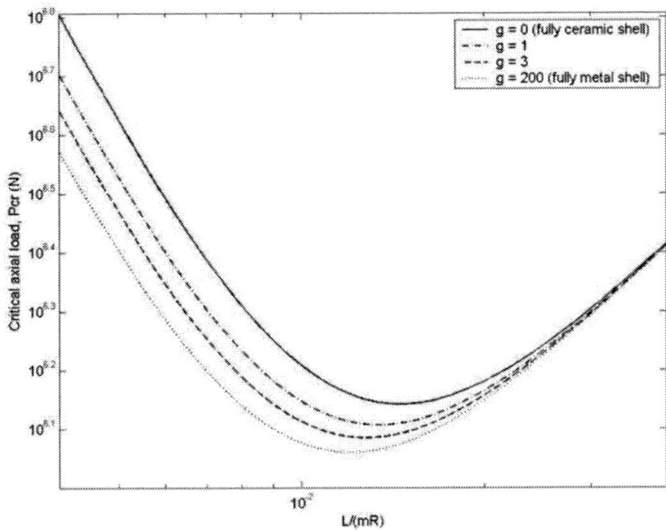

Figure 15. the variation of axial compressive buckling load of FGM cylindrical shell with an elastic layer versus L/mR for different values of index factor, g

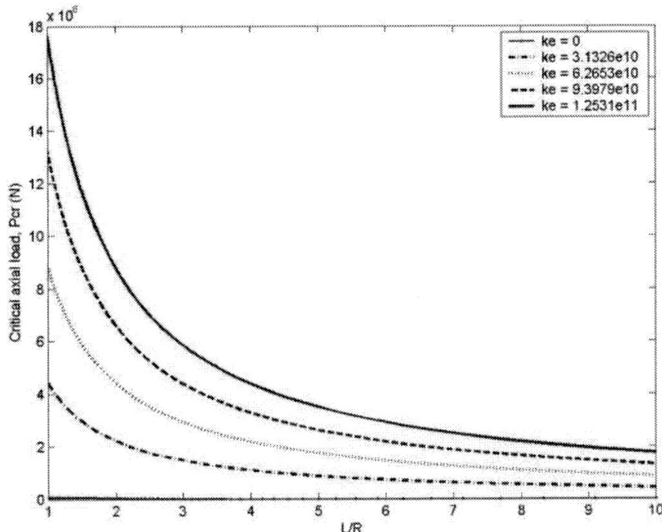

Figure 16. The Effect Of The Elastic Layer On The Buckling Behavior Of The FG Cylindrical Shell.

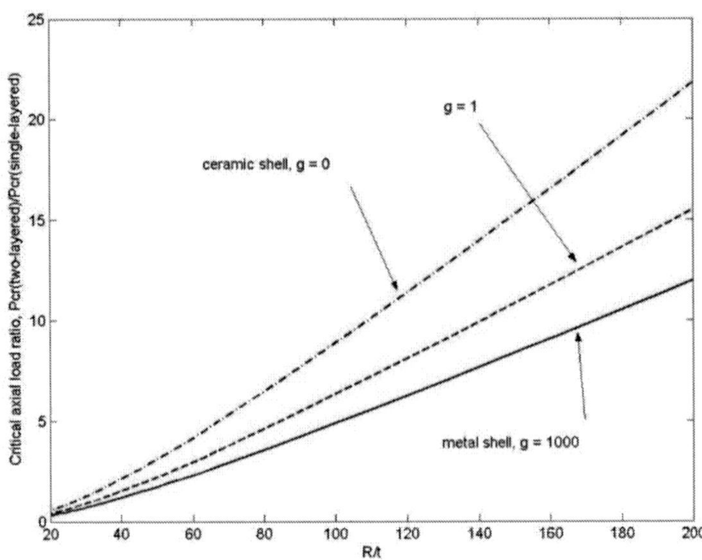

Figure 17. The ratio of elastic buckling load for uniaxial loading of a two-layered FGCS to that of a single layered FGCS with an equal mass and radius

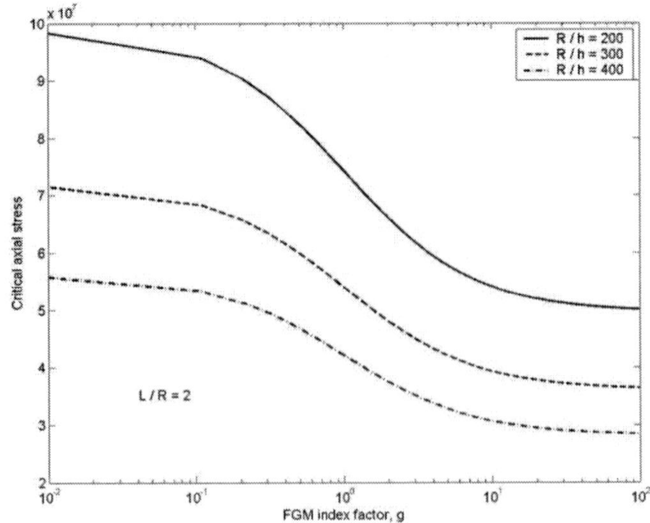

Figure 18. Effect of the FGM index factor on the critical axial stress of the FGCS.

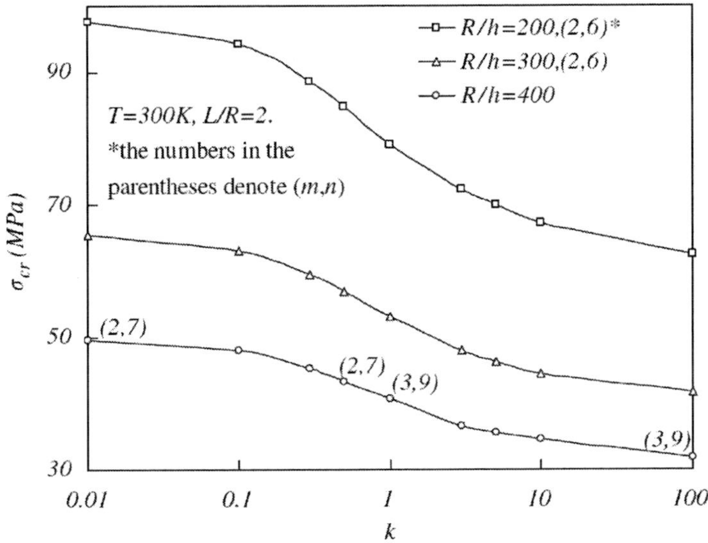

Figure 19. Effect of the FGM index factor on the critical axial stress of the FGCS (Huang and Han [23])

CONCLUSION

The buckling of a two-layered FG cylindrical shell subjected to axial force has been studied in this chapter. The assumed model in this study consists of one isotropic elastic cylindrical shell with functionally graded cylindrical core called two-layered FGCS. The new features of the effect of elastic layer on buckling of two-layered FG cylindrical shell and some meaningful results in this chapter are helpful for the application and the design of nuclear reactors, space planes and chemical plants, in which FG cylindrical layer act as basic element.

The elastic buckling of the two-layered FGCS has been analyzed, and its buckling resistance was compared with that of a FG hollow cylindrical shell. Results show that the critical axial load decreases with increasing FGM index factor. In addition, the application of an elastic layer increases elastic stability.

A comparison is made to the results obtained via a method based on FGM effective properties. It can be inferred from the results that the critical axial force of two-layered FGCS subjected to axial forces is dependent on the material composition and FGM index factor, and the shell geometry parameters and it is concluded that the application of an elastic layer increases elastic stability.

Material properties of functionally graded cylindrical shells are considered as temperature dependent and graded in the thickness direction according to a power-law distribution in terms of the volume fractions of the constituents. Finally, it is found that the effect of elastic layer on buckling of two-layered FGM cylindrical shell subjected to axial forces is not neglected in some cases and significantly reduces the weight of cylindrical shells.

REFERENCES

[1] Stein, M. The influence of prebuckling deformations and stresses on the buckling of perfect cylinders. NASA TR R-190, 1964.

[2] Shen HS, Chen TY. A boundary layer theory for the buckling of thin cylindrical shells under external pressure. *Applied Mathematics and Mechanics* 1988;9:557–71.

[3] Shen HS, Chen TY. A boundary layer theory for the buckling of thin cylindrical shells under axial compression. In: Chien WZ, Fu ZZ, editors. *Advances in Applied Mathematics and Mechanics in China*. Vol. 2.. Beijing, China: International Academic Publishers, 1990. p. 155–72.

[4] Shen HS, Chen TY. Buckling and postbuckling of cylindrical shells under combined external pressure and axial compression. *Thin-Walled Structures 1991*;12:321–34.

[5] Shen HS, Zhou P, Chen TY. Postbuckling analysis of stiffened cylindrical shells under combined external pressure and axial compression. *Thin-Walled Structures* 1993;15:43–63.

[6] Shen HS. Post-buckling analysis of imperfect stiffened laminated cylindrical shells under combined external pressure and axial compression. *Computers and Structures* 1997;63:335–48.

[7] Shen HS. Thermomechanical postbuckling analysis of stiffened laminated cylindrical shell. *Journal of Engineering Mechanics ASCE* 1997;123:433–43.

[8] Shen HS. Postbuckling analysis of imperfect stiffened laminated cylindrical shells under combined external pressure and thermal loading. *International Journal of Mechanical Sciences* 1998;40: 339–55.

[9] Shen HS. Postbuckling analysis of stiffened laminated cylindrical shells under combined external liquid pressure and axial compression. *Engineering Structures* 1998;20:738–51.

[10] Shen HS. Thermomechanical postbuckling of composite laminated cylindrical shells with local geometric imperfections. International *Journal of Solids and Structures* 1999;36(4):597–617.

[11] Karam, G. N. and Gibson, L. J., 1995, "Elastic Buckling of Cylindrical Shells with Elastic Cores I. Analysis," *Int. J. Solids Structures*, Vol. 32, No. 8/9, pp. 1259~1283.

[12] Agarwal, B. L. and Sobel, L. H., 1977, "Weight Comparisons of Optimized Stiffened, Unstiffened, and Sandwich Cylindrical Shells," *AIAA J.* 14, pp. 1000~1008.

[13] Hutchinson, J. W. and He M. Y., 2000, "Buckling of Cylindrical Sandwich Shells with Metal Foam Cores," Vol. 37 pp. 6777~6794.

[14] Brush, D. O. and Almroth, B. O., 1975, Buckling of Beams, Plates and Shells. McGraw-Hill, New York.

[15] Gough, G. J., Elam, C. F. and de Bruyne, N. A., 1940, "The Stabilization of a Thin Sheet by a Continuous Supporting Medium," *The Journal of Royal Aeronautical Society*, 44, pp. 12~43.

[16] Langhaar, H. L., 1962, "*Energy Methods in Applied Mechanics*. John Wiley," New York.

[17] Ng TY, Lam KM, Liew KM, Reddy JN. Dynamic stability analysis of functionally graded cylindrical shells under periodic axial loading. *International Journal of Solids and Structures* 2001; 38: 1295-309.

[18] Huaiwei Huang, Qiang Han, Nonlinear elastic buckling and postbuckling of axially compressed functionally graded cylindrical shells, *International Journal of Mechanical Sciences* 51 (2009) 500–507

Chapter 5

MAGNETOTHERMOELASTIC TRANSIENT RESPONSE OF A FUNCTIONALLY GRADED THICK HOLLOW SPHERE

ABSTRACT

In this chapter an analytical method is developed to obtain the response of magneto-thermo-elastic stress and perturbation of the magnetic field vector for a thick-walled spherical functionally graded materials (FGM) vessel. The vessel, which is placed in a uniform magnetic field, is subjected to an internal pressure and transient temperature gradient. Using the Hankel and Laplace transform techniques, the dynamic equation of magnetothermoelastic is solved and the radial and circumferential stresses as well as the perturbation of the magnetic field vector for a typical material are obtained. Moreover, the effect of magnetic field vector and material in-homogeneity on the stresses is investigated.

NOMENCLATURE

$\vec{U}, u_r$	displacement vector and radial displacement [m],
$\sigma_{rr}, \sigma_{\theta\theta}$	radial and circumferential stresses [N/m^2]
$\lambda_0, G_0, \gamma_0, \rho_0$ and μ_0	Lame's constant [N/m^2], shear modulus[N/m^2], thermal modulus[N/Km^2], mass density [kg/m^2] and

	magnetic permeability coefficient [H/m], respectively
m	material grading exponent
r	Radius [m]
$T(r,t)$	temperature change [K]
t, a, b time [s],	inner and outer radii of the FGM hollow sphere [m]
$\vec{H}, \vec{h}$	magnetic intensity vector, perturbation of magnetic field vector
$\vec{J}$	electric current density vector
$\vec{e}$	perturbation of electric field vector
$f_{\phi\phi}$	Lorentz's force [kg/m^2 s^2]
$P_i(t), P_o(t)$	internal and external pressure of FGM hollow sphere [N/m^2]
$u_{r0}(r), v_{r0}(r)$ I	nitial radial displacement [m], initial speed [m/s]
V_e	elastic wave speed of homogeneous sphere[m/s]
V_t	magnetothermoelastic wave speed [m/s]
ω	natural frequency of the FGM hollow sphere [1/s]

1. INTRODUCTION

As stated in previous chapters a functionally graded material is usually a combination of two material phases that has an intentional graded transition from one material at one surface to another material at the opposite surface. This transition allows the creation of multiple properties (or function) without any mechanically weak junction or interface. It is possible with these materials obtain a combination of properties that cannot be achieved in conventional monolithic materials. This makes FGMs suitable for many applications such as the nuclear reactor and high-speed space craft industries.

In recent years, a new area has been developed which studies the interactions between stress, strain, temperature (which is produced by the absorption of an electromagnetic pulse or γ ray pulse radiant energy) and electromagnetic fields. This new field is called the magneto-thermo-elasticity. The theory of magnetothermoelasticity has attracted the attention much attention due to its wide application in various fields such as geophysics for understanding the effect of the Earth's magnetic field on seismic waves, damping of acoustic waves in a magnetic field, emissions of electromagnetic radiations from nuclear devices, development of a highly sensitive superconducting magnetometer, electrical power engineering, optics, and plasma physics, to mention a few.

When a magnetic field acts on a conducting medium, the conducting medium is subjected to the Lorentz's force and heat supply of eddy current loss. Thus, two kinds of stress are generated: one is magnetic stress caused by the Lorentz's force and another is the thermal stress caused by eddy current loss. Many investigations have been conducted in the field of magnetoelasticity or magnetothermoelasticity [1]–[18]. However up to now, there have only been a few studies of the magnetothermoelastic behaviour of functionally gradient materials.

Nayfeh et al. [1] studied the propagation of plane waves in a solid under the influence of an electro-magnetic field. They obtained the governing equations for the general case and the solution for some particular cases. Choudhuri [2] extended these results for rotating media.

Dai et al. [3] studied the steady-state magnetoelastic behaviour of functionally graded materials cylindrical and spherical vessels placed in a uniform magnetic field and subjected to internal pressure analytically. Hosseini et al. [4] investigated the transient heat conduction in a cylindrical shell of functionally graded material by using analytical method.

Ootao et al. [5] analyzed the transient piezothermoelastic problem of a functionally graded thermo-piezoelectric hollow sphere due to a uniform heat supply using the Laplace transformation method.

Chand et al. [6] presented an investigation on the distribution of deformation, stresses and magnetic field in a homogeneous isotropic, thermally and electrically conducting, uniformly rotating elastic half-space.

Using the Laplace transform technique, Sherief et al.[7], obtained the distribution of thermal stresses and temperature in a generally thermoelastic electrically conducting half-space subjected to a rapid thermal shock and placed in a uniform magnetic field. Ezzat [8] obtained the distribution of thermal stresses and temperature in a perfectly conducting half-space which is suddenly heated to a constant temperature and compared the results with those obtained in the absence of a magnetic field.

Using the finite-difference method, Abd-Alla et al. [9],[10], solved the magnetothermoelastic problem of a nonhomogeneous isotropic cylinder in the presence of a heat source. Wang et al. [11] presented an analytical method to analyze the magneto–thermo–elastic waves and perturbation of the magnetic field vector produced by thermal shock in a solid conducting cylinder.

Dai et al. [12] presented an analytical solution for the interaction of electric potential, electric displacement, elastic deformations, mechanical loads, the described electro-magneto-elastic responses and perturbation of the magnetic field vector in a piezoelectric hollow cylinder subjected to rapid mechanical load and electric potential. Wang et al [13] presented a theoretical method for analyzing magnetothermoelastic responses and perturbation of the magnetic field vector in the conducting orthotropic thermo elastic cylinder subjected to thermal shock. They showed that the behavior of the magnetothermoestress-focusing effect and the perturbation response at the center of an orthotropic thermo elastic cylinder is different from that at the center of an orthotropic thermoelastic cylinder is different from that at the center of an isotropic thermo elastic cylinder.

Dai et al. [14] presented an analytical method to solve thermo-electro-elastic transient response in piezoelectric hollow structures subjected to arbitrary thermal shock, sudden mechanical load and electric excitation. They showed that the response histories and distributions of stresses, electric displacement and electric potential in a transversely isotropic piezoelectric hollow sphere are interaction each other.

Dai et al. [15] considered the steady-state magnetothermoelastic problem of a functionally graded material hollow structure subjected to mechanical loads. They assumed that the material properties obey the simple power-law variation through the structures' wall thickness. Using the infinitesimal theory of magnetothermoelasticity, they determined exact solutions for stresses and perturbations of the magnetic field vector in FGM hollow cylinders and FGM hollow spheres.

Wang et al. [16] presented a theoretical method for analyzing magnetothermoelastic response and perturbation of the magnetic field vector in a conducting non-homogeneous thermoelastic cylinder subjected to thermal shock. By making use of finite Hankle integral transforms, they obtained the analytical expressions for magntothermodynamic stress and perturbation response of an axial magnetic field vector in the non-homogeneous cylinder.

An analytical method to solve the problem for the dynamic stress-focusing and centered-effect of perturbation of the magnetic field vector in orthotropic cylinders under thermal and mechanical shock loads was introduced by Dai et al. [17]. They obtained analytical expressions for the dynamic stresses and the perturbation of the magnetic field vector by means of finite Hankel transforms and Laplace transforms.

Abd-Alla et al. [18] studied the generation of thermal stresses in a nonhomogeneous anisotropic solid cylinder rotating about the z-axis at a constant angular velocity in the presence of a magnetic field. They solved the governing equations numerically using the boundary-element method (BEM).

However, investigations of exact solutions for FGM hollow spheres, placed in a uniform magnetic field and transient temperature field, subjected to mechanical loads have not been found in the literature. In this chapter, an analytical method is developed to determine the response histories of stress and perturbation of the magnetic field vector in a functionally graded material hollow sphere placed in a uniform magnetic field, subjected to a sudden temperature change.

The medium is assumed traction free and without heat generation. The magneto–thermo–elastic equation of the functionally graded material hollow sphere is decomposed into a quasi-static homogeneous equation with inhomogeneous boundary conditions and an inhomogeneous dynamic equation with homogeneous boundary conditions. The quasi-static equation is first solved by direct integration. Next, the solution to the inhomogeneous dynamic equation which satisfies homogeneous boundary conditions is obtained utilizing the integral transforms, finite Hankel transforms [19] and the Laplace transforms techniques. Thus, the analytical solutions for the transient responses of displacements, stresses and perturbation of magnetic field vector in the FGM hollow sphere are obtained. Finally, numerical results are presented and discussed for the variation of the displacement, stresses and perturbation of magnetic field vector with the time and through the thickness of the FGM sphere.

2. Theoretical Formulation

A perfectly conducting functionally graded hollow sphere with internal radius a and external radius b placed in a circumferential magnetic field $\vec{H}(0,0,H_{\phi})$ is considered. It is assumed that the FGM hollow sphere is subjected to a rapid temperature change $T(r,t)$. Due to symmetry, we have $u_{\theta} = u_{\phi} = 0, u_r = u_r(r,t)$.

The stress-strain relations of the FGM hollow sphere are expressed as

$$\sigma_{rr}(r,t) = (\lambda + 2G)\frac{\partial u_r}{\partial r} + 2\lambda\frac{u_r}{r} - \gamma T(r,t) \tag{1a}$$

$$\sigma_{\theta\theta}(r,t) = 2(\lambda + G)\frac{u_r}{r} + \lambda\frac{\partial u_r}{\partial r} - \gamma T(r,t) \tag{1b}$$

where λ and G are the Lame's constants, and γ is the coefficient of linear thermal expansion.

The boundary conditions and initial conditions are

$$\sigma_{rr}(a,t) = -P_i(t)$$
$$\sigma_{rr}(b,t) = -P_o(t)$$

(2a)

$$u_r(r,0) = u_{r0}(r)$$
$$\left.\frac{\partial u_r(r,t)}{\partial t}\right|_{t=0} = v_{r0}(r) \tag{2b}$$

Assuming that the magnetic permeability, μ at the outer surface of the FGM sphere to be equal to the magnetic permeability of the medium around it, and the medium to be non-ferromagnetic and non-ferroelectric, and omitting the displacement electric currents, the governing electrodynamic Maxwell equations [20] for a perfectly conducting elastic body can be written as

$$\vec{J} = \nabla \times \vec{h}, \qquad \nabla \times \vec{e} = -\mu\frac{\partial \vec{h}}{\partial t}, \qquad div\vec{h} = 0,$$
$$\vec{e} = -\mu\left(\frac{\partial \vec{U}}{\partial t} \times \vec{H}\right), \qquad \vec{h} = Curl\left(\vec{U} \times \vec{H}\right) \tag{3}$$

Applying a primary magnetic field vector $\vec{H}(0,0,H_\phi)$ to Eq. (3), yields

$$\vec{U} = (u_r, 0, 0), \quad \vec{e} = -\mu\left(0, H_\varphi \frac{\partial u_r}{\partial t}, 0\right)$$
$$\vec{h} = (0, 0, h_\varphi), \vec{J} = \left(0, -\frac{\partial h_\varphi}{\partial r}, 0\right), \; h_\varphi = -H_\varphi\left(\frac{\partial u_r}{\partial r} + \frac{2u_r}{r}\right) \tag{4}$$

$f_{\phi\phi}$ is defined as the Lorentz's force [20], which may be written as

$$f_{\phi\phi} = \mu(r)(\vec{J} \times \vec{H}) = \mu(r) H_\phi^2 \frac{\partial}{\partial r}\left(\frac{\partial u_r}{\partial r} + \frac{2u_r}{r}\right) \tag{5}$$

In the absence of external body forces and inner heat sources, and taking into account the Lorentz's force, the equation of motion for a FGM hollow sphere is expressed as

$$\frac{\partial \sigma_{rr}}{\partial r} + \frac{2(\sigma_{rr} - \sigma_{\theta\theta})}{r} + f_{\phi\phi} = \rho \frac{\partial^2 u_r}{\partial t^2} \qquad a \le r \le b \;, \quad t \ge 0 \tag{6}$$

where ρ is the mass density.

The in-homogeneity in the material is assumed to vary according to the following relations

$$\bar{r} = {r}/{b}, \; \lambda(r) = \lambda_0 \bar{r}^{2m}, G(r) = G_0 \bar{r}^{2m}, \gamma(r) = \gamma_0 \bar{r}^{2m}, \mu(r) = \mu_0 \bar{r}^{2m}, \rho(r) = \rho_0 \bar{r}^{2m} \tag{7}$$

Substituting Eq. (7) into Eqs. (1) yields

$$\sigma_{rr}(r,t) = \bar{r}^{2m}[(\lambda_0 + 2G_0)\frac{\partial u_r}{\partial r} + 2\lambda_0 \frac{u_r}{r} - \gamma_0 T(r,t)] \tag{8a}$$

$$\sigma_{\theta\theta}(r,t) = \bar{r}^{2m}[2(\lambda_0 + G_0)\frac{u_r}{r} + \lambda_0 \frac{\partial u_r}{\partial r} - \gamma_0 T(r,t)] \tag{8b}$$

For the analysis, the following dimensionless quantities are introduced

$$u = \frac{u_r}{b} \ , \ \sigma_r(r,t) = \frac{\sigma_{rr}(r,t)}{\lambda_0 + 2G_0} \ , \ \sigma_\theta(r,t) = \frac{\sigma_{\theta\theta}(r,t)}{\lambda_0 + 2G_0}$$

$$D_1 = \frac{2\lambda_0}{\lambda_0 + 2G_0} \ , \ D_2 = \frac{2(\lambda_0 + G_0)}{\lambda_0 + 2G_0} \ , \ D_3 = \frac{\mu_0 H_\phi^2}{\lambda_0 + 2G_0}$$

$$\tau = \frac{V_e t}{b} \ , \ V_e = \sqrt{\frac{\lambda_0 + 2G_0}{\rho_0}} \ , \ \bar{T}(\bar{r},\tau) = \frac{\gamma_0}{\lambda_0 + 2G_0} T(r,t)$$

$$f_\phi = \frac{f_{\phi\phi}}{\lambda_0 + 2G_0} b \ , \ q = \frac{a}{b} \ , \ \bar{P}_i(\tau) = \frac{P_i(t)}{\lambda_0 + 2G_0} \ , \ \bar{P}_o(\tau) = \frac{P_o(t)}{\lambda_0 + 2G_0} \tag{9}$$

Using the above dimensionless variables, the Eqs. (8), (4) and (6) can be expressed as

$$\sigma_r(\bar{r},\tau) = \bar{r}^{2m}\left[\frac{\partial u}{\partial \bar{r}} + D_1 \frac{u}{\bar{r}} - \bar{T}(\bar{r},\tau)\right] \tag{10a}$$

$$\sigma_\theta(\bar{r},\tau) = \bar{r}^{2m}\left[D_2 \frac{u}{\bar{r}} + \frac{D_1}{2}\frac{\partial u}{\partial \bar{r}} - \bar{T}(\bar{r},\tau)\right] \tag{10b}$$

$$h_\phi(\bar{r},\tau) = -H_\phi\left(\frac{\partial u}{\partial \bar{r}} + \frac{2u}{\bar{r}}\right) \tag{11}$$

$$\frac{\partial \sigma_r}{\partial \bar{r}} + \frac{2(\sigma_r - \sigma_\theta)}{\bar{r}} + f_\phi = \bar{r}^{2m}\frac{\partial^2 u}{\partial \tau^2} \qquad q \le \bar{r} \le 1 \ , \quad \tau \ge 0 \tag{12}$$

Substituting Eqs. (10) and (5) into Eq. (12) and using Eq. (9) yield the following displacement equilibrium equation of magnetothermoelastic in the FGM hollow sphere

$$\frac{\partial^2 u}{\partial \bar{r}^2} + \frac{2(1+N)}{\bar{r}}\frac{\partial u}{\partial \bar{r}} - \frac{M_t^2}{\bar{r}^2} u = \frac{1}{V_t^2}\frac{\partial^2 u}{\partial \tau^2} + S(\bar{r},\tau) \qquad q \le \bar{r} \le 1 \tag{13}$$

where:

$$N = \frac{m}{1+D_3} \ , \ V_t^2 = 1 + D_3 \ , \ M_t^2 = \frac{2(D_2 + D_3) - D_1(1+2m)}{1+D_3} \tag{14}$$
$$S(\bar{r},\tau) = \frac{2m}{\bar{r}V_t^2}\bar{T}(\bar{r},\tau) + \frac{1}{V_t^2}\frac{\partial \bar{T}}{\partial \bar{r}}$$

According to the previous equations, the boundary conditions (2a) and the initial condition (2b) are rewritten as

$$q^{2m}\left(\left.\frac{\partial u}{\partial \bar{r}}\right|_{\bar{r}=q} + D_1\frac{u}{q} - \bar{T}(q,\tau)\right) = -\bar{P}_i(\tau)$$
$$\left.\frac{\partial u}{\partial \bar{r}}\right|_{\bar{r}=1} + D_1 u - \bar{T}(1,\tau) = -\bar{P}_o(\tau) \tag{15a}$$

$$u(\bar{r},0) = u_0(\bar{r})$$
$$\left.\frac{\partial u(\bar{r},\tau)}{\partial \tau}\right|_{\tau=0} = v_0(\bar{r}) \tag{15b}$$

A new variable $y(\bar{r},\tau)$ is introduced in order to convert Eq. (13) into a normal Bessel equation.

$$u(\bar{r},\tau) = \bar{r}^{-(N+\frac{1}{2})}\, y(\bar{r},\tau) \tag{16}$$

Thus, Eq. (13) is rewritten as

$$\frac{\partial^2 y}{\partial \bar{r}^2} + \frac{1}{\bar{r}}\frac{\partial y}{\partial \bar{r}} - \frac{M^2}{\bar{r}^2}y = \frac{1}{V_t^2}\frac{\partial^2 y}{\partial \tau^2} + \bar{r}^{N+\frac{1}{2}}S(\bar{r},\tau) \quad q \le \bar{r} \le 1 \tag{17}$$

where:

$$M^2 = M_t^2 + N^2 + N + \frac{1}{4} \tag{18a}$$

Using Eq. (16), the boundary conditions and initial conditions (15) are written as

$$\left.\frac{\partial y}{\partial \bar{r}}\right|_{\bar{r}=q} + g\frac{y}{q} = q^{N+\frac{1}{2}}\bar{T}(q,\tau) - q^{N-2m+\frac{1}{2}}\bar{P}_i(\tau) = \Theta_1(\tau)$$
$$\left.\frac{\partial y}{\partial \bar{r}}\right|_{\bar{r}=1} + gy = \bar{T}(1,\tau) - \bar{P}_o(\tau) = \Theta_2(\tau) \qquad (18b)$$

$$y(\bar{r},0) = y_0(\bar{r})$$
$$\left.\frac{\partial y(\bar{r},\tau)}{\partial \tau}\right|_{\tau=0} = y_0'(\bar{r}) \qquad (18c)$$

where

$$g = D_1 - N - \frac{1}{2} \qquad (18d)$$

2.1. Solution Method

The following general solution is assumed for Eq. (17)

$$y(\bar{r},\tau) = y_q(\bar{r},\tau) + y_d(\bar{r},\tau) \qquad (19)$$

where $y_q(\bar{r},\tau)$ and $y_d(\bar{r},\tau)$ are the quasi-static solution which satisfies inhomogeneous boundary conditions, and dynamic solution which satisfies homogeneous boundary conditions, respectively.

2.1.1. Quasi-Static Solution

In this section, the quasi-static solution of the Eq. (17) is derived. Disregarding the inertia term of the right-hand side in Eq. (17), the quasi-static solution $y_q(\bar{r},\tau)$ must satisfy the following equation and inhomogeneous boundary condition

$$\frac{\partial^2 y_q}{\partial \bar{r}^2} + \frac{1}{\bar{r}}\frac{\partial y_q}{\partial \bar{r}} - \frac{M^2}{\bar{r}^2} y_q = \bar{r}^{N+\frac{1}{2}} S(\bar{r},\tau) \qquad q \le \bar{r} \le 1 \tag{20}$$

$$\left.\frac{\partial y_q}{\partial \bar{r}}\right|_{\bar{r}=q} + g \left.\frac{y_q}{q}\right|_{\bar{r}=q} = \Theta_1(\tau)$$
$$\left.\frac{\partial y_q}{\partial \bar{r}}\right|_{\bar{r}=1} + g y_q\Big|_{\bar{r}=1} = \Theta_2(\tau) \tag{21a}$$

Equation (20) can be simplified to

$$\frac{\partial}{\partial \bar{r}}\left[\bar{r}^{1-2M} \frac{\partial}{\partial \bar{r}}\left(\bar{r}^{M} y_q(\bar{r},\tau)\right)\right] = \bar{r}^{N-M+\frac{3}{2}} S(\bar{r},\tau) \tag{21b}$$

The quasi-static solution for Eq. (20), which satisfies the boundary condition (21a), is expressed as

$$y_q(\bar{r},\tau) = \bar{r}^{-M} \int_q^{\bar{r}} \bar{r}^{2M-1} \left(\int_q^{\bar{r}} \bar{r}^{N-M+\frac{3}{2}} S(\bar{r},\tau) d\bar{r}\right) d\bar{r} + \frac{A_1}{2M}\left(\bar{r}^{M} - \bar{r}^{-M} q^{2M}\right) + A_2 \bar{r}^{-M} \tag{22}$$

where

$$M = \sqrt{M_t^2 + N^2 + N + 1/4} \tag{23a}$$

Using the boundary condition (21a), the constants A_1 and A_2 in Eq. (22) are determined

$$A_1 = \frac{\Theta_2(\tau) - \Theta_1(\tau) q^{M+1} - \int_q^1 \bar{r}^{N-M+\frac{3}{2}} S(\bar{r},\tau) d\bar{r} + (M-g)\int_q^1 \bar{r}^{2M-1} \left[\int_q^{\bar{r}} \bar{r}^{N-M+\frac{3}{2}} S(\bar{r},\tau) d\bar{r}\right] d\bar{r}}{\frac{M+g}{2M}(1-q^{2M})} \tag{23b}$$

$$A_2 = \left(\Theta_1(\tau)(\frac{M+g}{2M} q^{M+1} + \frac{M-g}{2M} q^{3M+1}) - q^{2M}\Theta_2(\tau) + q^{2M}\int_q^1 \bar{r}^{N-M+\frac{3}{2}} S(\bar{r},\tau)d\bar{r} \right.$$
$$\left. + (g-M)q^{2M}\int_q^1 \bar{r}^{2M-1}[\int_q^{\bar{r}} \bar{r}^{N-M+\frac{3}{2}} S(\bar{r},\tau)d\bar{r}]d\bar{r} \right) \Big/ \left(\frac{g^2-M^2}{2M}(1-q^{2M}) \right) \quad (23c)$$

2.1.2. Dynamic Solution

The dynamic solution $y_d(\bar{r},\tau)$ should satisfy the following inhomogeneous equation (24), the corresponding homogeneous boundary conditions (25a) and initial condition (25b).

$$\frac{\partial^2 y_d}{\partial \bar{r}^2} + \frac{1}{\bar{r}}\frac{\partial y_d}{\partial \bar{r}} - \frac{M^2}{\bar{r}^2} y_d = \frac{1}{V_t^2}(\frac{\partial^2 y_q}{\partial \tau^2} + \frac{\partial^2 y_d}{\partial \tau^2}) \quad q \le \bar{r} \le 1 \quad (24)$$

$$\left.\frac{\partial y_d}{\partial \bar{r}}\right|_{\bar{r}=q} + g\left.\frac{y_d}{q}\right|_{\bar{r}=q} = 0$$
$$\left.\frac{\partial y_d}{\partial \bar{r}}\right|_{\bar{r}=1} + g\left. y_d\right|_{\bar{r}=1} = 0 \quad (25a)$$

$$y_q(\bar{r},0) + y_d(\bar{r},0) = y_0(\bar{r})$$
$$\left.\frac{\partial y_q(\bar{r},\tau)}{\partial \tau}\right|_{\tau=0} + \left.\frac{\partial y_d(\bar{r},\tau)}{\partial \tau}\right|_{\tau=0} = y_0'(\bar{r}) \quad (25b)$$

The solution of homogeneous part of Eq. (24) (assuming $y_q(\bar{r},\tau)=0$) is assumed to be

$$y_{d0}(\bar{r},\tau) = Z(\bar{r})e^{i\omega\tau} \quad (26)$$

where $Z(\bar{r})$ and ω are the characteristic function and natural frequency, respectively.

Substituting Eq. (26) into the homogeneous part of Eq. (24) and Eq. (25a) yields

$$\frac{d^2 Z(\bar{r})}{d\bar{r}^2}+\frac{1}{\bar{r}}\frac{dZ(\bar{r})}{d\bar{r}}+(k_i{}^2-\frac{M^2}{\bar{r}^2})Z(\bar{r})=0 \quad , \ \omega_i = V_t k_i \tag{27}$$

$$\left.\frac{dZ(\bar{r})}{d\bar{r}}\right|_{\bar{r}=q}+g\frac{Z(q)}{q}=0$$
$$\left.\frac{dZ(\bar{r})}{d\bar{r}}\right|_{\bar{r}=1}+gZ(1)=0 \tag{28}$$

Equation (27) is a Bessel equation and its general solution is given by

$$Z(\bar{r})=B_1 J_M(k_i\bar{r})+B_2 Y_M(k_i\bar{r}) \tag{29}$$

where B_1 and B_2 are unknown constants which can be determined by using of the boundary condition (28) as follows

$$B_1\chi_a + B_2\beta_a = 0$$
$$B_1\chi_b + B_2\beta_b = 0 \tag{30}$$

where χ_a, χ_b, β_a and β_b are expressed as

$$\chi_a = k_i J'_M(k_i q)+\frac{g}{q}J_M(k_i q), \tag{31a}$$

$$\chi_b = k_i J'_M(k_i)+gJ_M(k_i), \tag{31b}$$

$$\beta_a = k_i Y'_M(k_i q)+\frac{g}{q}Y_M(k_i q), \tag{31c}$$

$$\beta_b = k_i Y'_M(k_i)+gY_M(k_i). \tag{31d}$$

Thus, the corresponding eigen-equation is expressed as

$$\chi_a \beta_b - \chi_b \beta_a = 0 \tag{32}$$

$J_M(k_i\bar{r})$ and $Y_M(k_i\bar{r})$ are the M-th order Bessel function of the first kind and second kind, respectively, k_i $(i = 1,2,..., \infty)$ are the positive roots of the eigen-equation (32).

The natural frequencies are, then, given by

$$\omega_i = V_t k_i \tag{33}$$

Next, the following Hankel transformation is introduced

$$\bar{Z}(k_i) = H[Z(\bar{r})] = \int_q^1 \bar{r} Z(\bar{r})\, N_M(k_i\bar{r})\, d\bar{r} \tag{34}$$

The inverse transformation is

$$Z(\bar{r}) = \sum_{k_i}^{\infty} \frac{\bar{Z}(k_i)}{\|N_M(k_i)\|} N_M(k_i\bar{r}) \tag{35}$$

where

$$N_M(k_i\bar{r}) = J_M(k_i\bar{r})\beta_a - \chi_a Y_M(k_i\bar{r}), \tag{36a}$$

$$\|N_M(k_i)\| = \int_q^1 \bar{r}[N_M(k_i\bar{r})]^2 d\bar{r}$$
$$= \frac{2\chi_a^2}{(\pi k_i \chi_b)^2}[g^2 + k_i^2(1-(\frac{M}{k_i})^2)] - \frac{2}{(\pi k_i)^2}[(\frac{g}{q})^2 + k_i^2(1-(\frac{M}{qk_i})^2)]. \tag{36b}$$

Hankel transforming Eq. (24) yields

$$N_M(k_i)(y_d'(1,\tau)+gy_d(1,\tau))-k_i^{\,2}\bar{y}_d(k_i,\tau)=\frac{1}{V_t^2}\left(\frac{\partial^2\bar{y}_q}{\partial\tau^2}+\frac{\partial^2\bar{y}_d}{\partial\tau^2}\right) \tag{37}$$

Using the homogeneous boundary condition (25a), the first term on the left-hand side of the above equation reduces to zero. Therefore, Equation (37) simplifies to the following equation

$$-k_i^{\,2}\bar{y}_d(k_i,\tau)=\frac{1}{V_t^2}\left(\frac{\partial^2\bar{y}_q}{\partial\tau^2}+\frac{\partial^2\bar{y}_d}{\partial\tau^2}\right) \tag{38}$$

Taking Laplace transform of both sides of Eq. (38) and using the initial condition (25b) yield

$$-k_i^{\,2}V_t^{\,2}\bar{y}_d^*(k_i,s)=s^2\bar{y}_d^*(k_i,s)+s^2\bar{y}_q^*(k_i,s)-s\bar{y}_0(k_i)-\bar{y}_0'(k_i) \tag{39}$$

Equation (39) can be simplified to

$$\bar{y}_d^*(k_i,s)=\omega_i\frac{\omega_i}{\omega_i^{\,2}+s^2}\bar{y}_q^*(k_i,s)-\bar{y}_q^*(k_i,s)+\frac{s}{\omega_i^{\,2}+s^2}\bar{y}_0(k_i)+\frac{\bar{y}_0'(k_i)}{\omega_i^{\,2}+s^2} \tag{40}$$

where $\bar{y}_0(k_i)=H[y_0(\bar{r})]$ and $\bar{y}_0'(k_i)=H[y_0'(\bar{r})]$. Taking inverse transform of Eq. (40) yields

$$\begin{aligned}\bar{y}_d(k_i,\tau)=\omega_i\int_0^\tau\bar{y}_q(k_i,t)\sin(\omega_i(\tau-t))dt-\bar{y}_q(k_i,\tau)+\\+\bar{y}_0(k_i)\cos(\omega_i\tau)+\frac{\bar{y}_0'(k_i)}{\omega_i}\sin(\omega_i\tau)\end{aligned} \tag{41}$$

Thereafter, in the numerical results section, we will assume

$$y_0(\bar{r})=0,\; y_0'(\bar{r})=0.$$

Then $y_d(\bar{r},\tau)$ can be obtained from

$$y_d(\bar{r},\tau)=\sum_{k_i}^{\infty}\frac{\bar{y}_d(k_i,\tau)}{\|N_M(k_i)\|}N_M(k_i\bar{r}) \tag{42}$$

2.1.3. General Solution

Substituting Eq.(42) into Eqs. (19) and (16), the solution of the basic displacement equation of magnetothermoelastic motion in the FGM hollow sphere is expressed as

$$u(\bar{r},\tau)=\bar{r}^{-(N+\frac{1}{2})}\Bigg[\bar{r}^{-M}\int_q^{\bar{r}}\bar{r}^{2M-1}\Big(\int_q^{\bar{r}}\bar{r}^{N-M+\frac{3}{2}}S(\bar{r},\tau)d\bar{r}\Big)d\bar{r}+\frac{A_1}{2M}(\bar{r}^{M}-\bar{r}^{-M}q^{2M})+A_2\bar{r}^{-M}$$
$$+\sum_{k_i}^{\infty}\frac{\bar{y}_d(k_i,\tau)}{\|N_M(k_i)\|}N_M(k_i\bar{r})\Bigg] \tag{43}$$

Substituting Eq. (43) into Eqs. (10) and (11), the radial stress, circumferential stress and the perturbation of magnetic field vector can be written as

$$\sigma_r(\bar{r},\tau)=g\bar{r}^{2m-1}u(\bar{r},\tau)-M\bar{r}^{2m-M-N-\frac{3}{2}}\int_q^{\bar{r}}\bar{r}^{2M-1}\Big(\int_q^{\bar{r}}\bar{r}^{N-M+\frac{3}{2}}S(\bar{r},\tau)d\bar{r}\Big)d\bar{r}$$
$$+\bar{r}^{2m+M-N-\frac{3}{2}}\int_q^{\bar{r}}\bar{r}^{N-M+\frac{3}{2}}S(\bar{r},\tau)d\bar{r}+\frac{A_1}{2}\Big(\bar{r}^{2m+M-N-\frac{3}{2}}+q^{2M}\bar{r}^{2m-M-N-\frac{3}{2}}\Big)$$
$$-A_2M\bar{r}^{2m-M-N-\frac{3}{2}}+\bar{r}^{2m-N-\frac{1}{2}}\sum_{k_i}^{\infty}k_i\frac{\bar{y}_d(k_i,\tau)}{\|N_M(k_i)\|}(J'_M(k_i\bar{r})\beta_a-\chi_aY'_M(k_i\bar{r}))$$
$$-\bar{r}^{2m}\bar{T}(\bar{r},\tau) \tag{44a}$$

$$\sigma_\theta(\bar{r},\tau) = (-\frac{D_1}{2}(N+\frac{1}{2}) + D_2)\bar{r}^{2m-1}u(\bar{r},\tau) - \frac{D_1 M}{2}\bar{r}^{2m-M-N-\frac{3}{2}}\int_q^{\bar{r}} \bar{r}^{2M-1}(\int_q^{\bar{r}} \bar{r}^{N-M+\frac{3}{2}} S(\bar{r},\tau)d\bar{r})d\bar{r}$$
$$+\frac{D_1}{2}\bar{r}^{2m+M-N-\frac{3}{2}}\int_q^{\bar{r}} \bar{r}^{N-M+\frac{3}{2}} S(\bar{r},\tau)d\bar{r} + \frac{A_1 D_1}{4}(\bar{r}^{2m+M-N-\frac{3}{2}} + q^{2M}\bar{r}^{2m-M-N-\frac{3}{2}})$$
$$-\frac{MA_2 D_1}{2}\bar{r}^{2m-M-N-\frac{3}{2}} + \frac{D_1}{2}\bar{r}^{2m-N-\frac{1}{2}}\sum_{k_i}^{\infty} k_i \frac{\bar{y}_d(k_i,\tau)}{\|N_M(k_i)\|}(J'_M(k_i\bar{r})\beta_a - \chi_a Y'_M(k_i\bar{r}))$$
$$-\bar{r}^{2m}\bar{T}(\bar{r},\tau) \tag{44b}$$

$$h_\phi(\bar{r},\tau) = -H_\phi[\frac{-N+3/2}{\bar{r}}u(\bar{r},\tau) - M\bar{r}^{-M-N-3/2}\int_q^{\bar{r}} \bar{r}^{2M-1}(\int_q^{\bar{r}} \bar{r}^{N-M+\frac{3}{2}} S(\bar{r},\tau)d\bar{r})d\bar{r}$$
$$+\bar{r}^{M-N-3/2}\int_q^{\bar{r}} \bar{r}^{N-M+\frac{3}{2}} S(\bar{r},\tau)d\bar{r} + \frac{A_1}{2}(\bar{r}^{M-N-\frac{3}{2}} + q^{2M}\bar{r}^{-M-N-\frac{3}{2}}) - A_2 M\bar{r}^{-M-N-\frac{3}{2}}$$
$$+\bar{r}^{-(N+\frac{1}{2})}\sum_{k_i}^{\infty} k_i \frac{\bar{y}_d(k_i,\tau)}{\|N_M(k_i)\|}(J'_M(k_i\bar{r})\beta_a - \chi_a Y'_M(k_i\bar{r}))] \tag{45}$$

NUMERICAL RESULTS AND DISCUSSION

In the previous section the transient response of a functionally graded material hollow sphere placed in a uniform magnetic field and subjected to a sudden temperature change has been obtained.

Table 1. The first five values of natural frequency of FGM sphere for various magnetic intensity vector

	$\omega_1(1/s)$	$\omega_2(1/s)$	$\omega_3(1/s)$	$\omega_4(1/s)$	$\omega_5(1/s)$
$H_\varphi = 2.23\times10^9$	16.024	362.786	725.090	1087.501	1449.942
$H_\varphi = 10^8$	1.560	35.395	70.712	106.046	141.384
$H_\varphi = 0$	1.386	31.471	62.860	94.266	125.678

The general solution of the magnetothermoelastic equation (13) which has two parts, namely quasi-static and dynamic solutions, has been presented. Disregarding the inertia term in Eq. (17), the quasi-static solution which satisfies the homogeneous differential equation with non-homogeneous boundary conditions, is obtained. The solution to the non-homogeneous equation with homogeneous boundary conditions, which is associated with radial vibration and propagation of magnetothermoelastic waves in the FGM sphere, constitutes the dynamic part.

The exact expressions for transient responses of displacements, stresses, and perturbation of magnetic field vector in the FGM hollow sphere are obtained by substituting Eq.(43) into Eqs. (10) and (11). The following, results of the displacement distribution, stresses and perturbation of the magnetic field vector are presented.

A transitory temperature change produced by a sudden electric current pulse or by absorption of electromagnetic wave with a life much less than 1 μs which varies along the radial direction of the FGM sphere, is considered

$$T(\bar{r},\tau) = \bar{r}^{\,m} T_0 H(\tau) \tag{46}$$

where $H(\tau)$ is the Heaviside function that may be expressed as

$$H(\tau) = \begin{cases} 0 & \tau < 0 \\ NAN & \tau = 0 \\ 1 & \tau > 0 \end{cases} \tag{47}$$

The following material constants for the FGM hollow sphere are employed for the calculation [15]

$$\lambda_0 = 28.8Gpa,\ \ G_0 = 9.25Gpa,\ \ \gamma_0 = 1.43(Mpa/C),\ \ \rho_0 = 1743(kg/m^3),$$
$$\mu_0 = 4\pi\times10^{-7}(H/m),\ \ H_\phi = 2.23\times10^9(A/m). \tag{48}$$

The inner and outer radius of the sphere is assumed to be 0.9m and 1m, respectively. The boundary conditions at the inner and outer surface of the sphere are assumed to be $P_i(t) = 100Mpa, P_o(t) = 0$, respectively.

In order to insure accuracy of the results, thirty eigen values have been used in the numerical calculations. Figures 1 show the history of the radial magneto-thermo-elastic stresses for different radius of the FGM spherical vessel with inhomogeneity constant $m = 3$. It is observed from these figures that the radial stresses at the inner and outer surface of the vessel are constant and satisfy the imposed boundary conditions. Moreover, it is noticed that the radial stresses oscillate around the quasi-static solution of the problem. The oscillation is due to reflected waves from the outer surface of the sphere. It is seen from the figure that the reaction of magnetothermoelastic waves is different at different points along the radius which is due to the material inhomogeneity. Moreover, the reaction time for maximum radial stress differs at different points along the radius. For example, at $\bar{r} = 0.91$, the reaction of maximum radial stress occurs at $\tau = 2.6$ and for $\bar{r} = 0.93$, 0.95 and 0.99 it occurs at $\tau = 6.4$, $\tau = 4.2$ and $\tau = 0.2$, respectively.

Figures 2 show the history of reaction of circumferential Magnetothermoelastic stresses at various points along the radius. It is observed from Figs. 2 that the circumferential Magnetothermoelastic stresses oscillate around the quasi-static solution of the problem. In order to avoid the effect of reflected waves, the results are presented up to $\tau = 5$. The results show that the shape of circumferential stress waves is completely different from that of radial stress wave. The amplitude of these waves decreases with time and eventually they tend to the quasi-static solution of the problem. Moreover, the amplitude of circumferential magnetothermoelastic stress waves and the constant value of the stress increase with increasing the radius of sphere. The maximum of the stress occurs at the outer surface of the vessel. Moreover, the circumferential stress waves are always tensile which reduces the fatigue life of FGM sphere subjected to a sudden temperature gradient and uniform magnetic field. The response histories of perturbation of magnetic vector fields for different points along the radius are depicted in Figs. 3.

These waves, similar to circumferential stress waves are symmetric and oscillate around the quasi-static solution of the problem and their amplitudes decreases with time. Eventually, they tend to the quasi-static solution of the problem. Contrary to the results of Figs. 2, the amplitude of these waves decreases with increasing the radius. The five first frequencies of the FGM spherical vessel with $m = 1$ for different intensities of the magnetic field are presented in table 1. As can be seen, the frequencies decrease with decreasing H_φ, which results in a reduction of the speed of magneto-thermo-elastic waves.

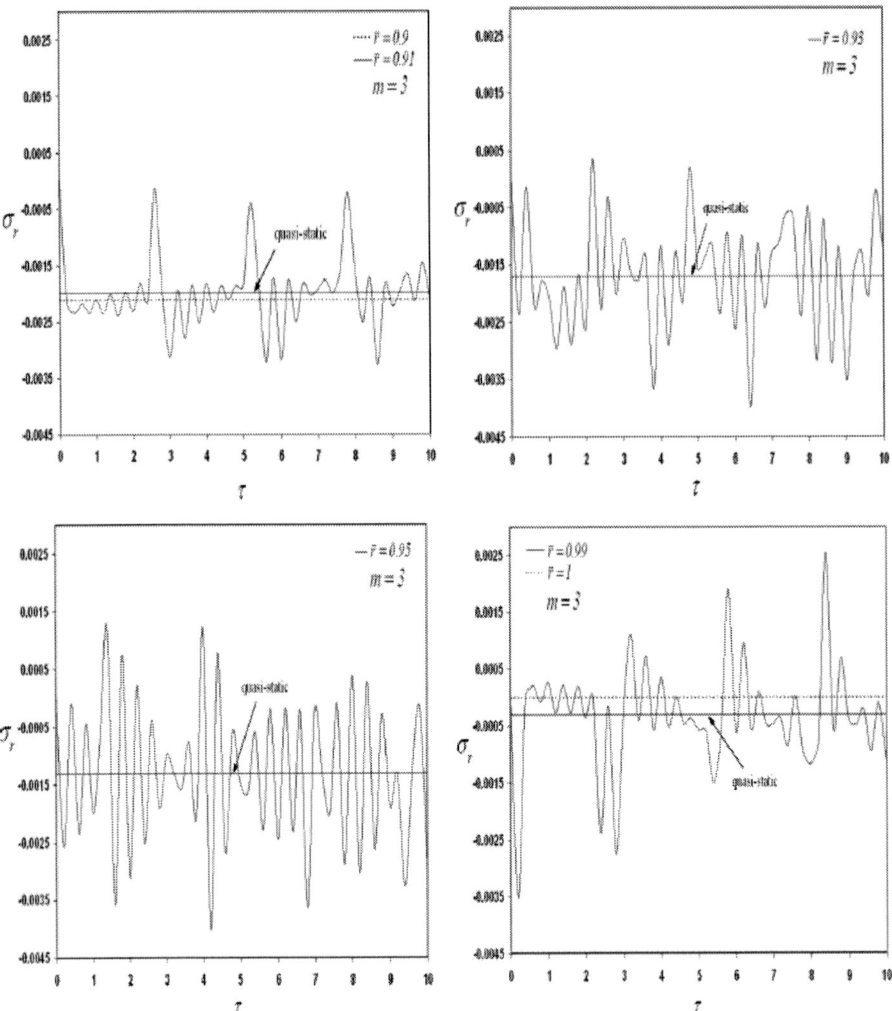

Figure 1. Response histories of radial stress in a functionally graded hollow sphere subjected to an internal pressure and thermal shock placed in a uniform circumferential magnetic field, where $\bar{r} = r/b$, $b = 1\,m$ and $\tau = V_e t/b$.

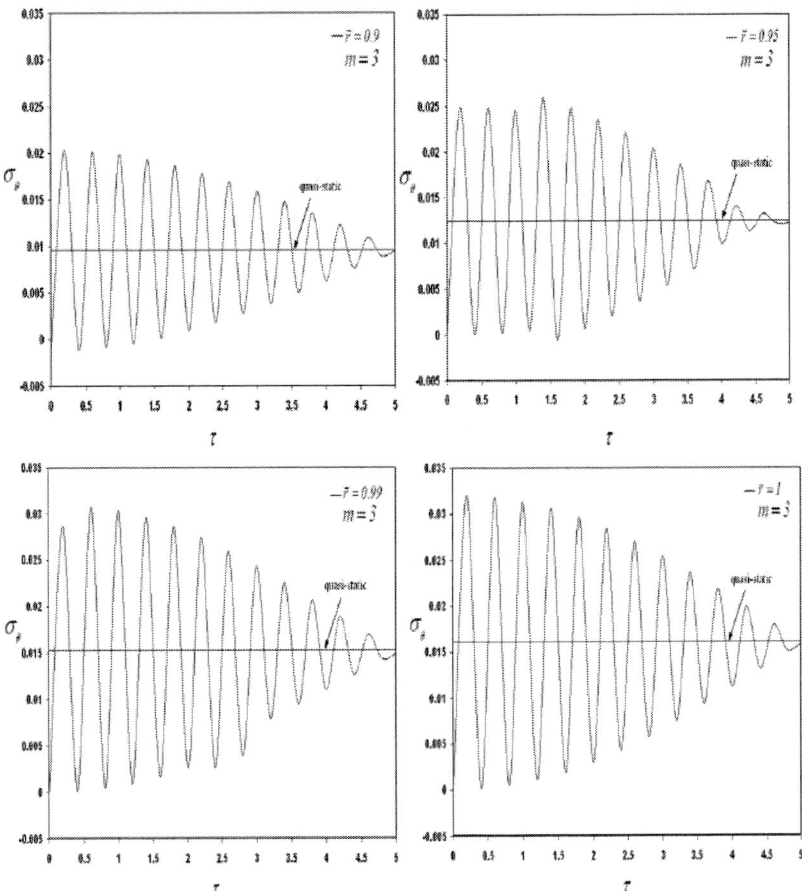

Figure 2. Response histories of circumferential stress in a functionally graded hollow sphere subjected to an internal pressure and thermal shock placed in a uniform circumferential m agnetic field, where $\bar{r} = r/b$, $b = 1m$ and $\tau = V_e t/b$.

The results can be seen in Figs. 4 through 6. In these figures, the effect of intensity of the magnetic field vector on the transient responses of the FGM spherical pressure vessel is depicted. As can be seen, the speed of response of the waves reduces drastically with decreasing H_{φ}. Figures 7 through 9 show the transient response of radial and circumferential stresses as well as perturbation of the magnetic field vector for the FGM sphere subjected to transient temperature gradient and uniform magnetic field for different values of inhomogeneity constant at $\bar{r} = 0.95$. $m = 0$ is for homogeneous material.

As can be seen from the figures, the shape of the waves remains unchanged with increasing m . However, the maximum amplitude of the waves changes with increasing the non-homogeneity constant.

It is clear from Figure 7 that the absolute value of the radial stress increases with increasing the non-homogeneity constant. Similar trends are observed in Figs. 8 and 9.

Figures 10 and 11 show the response history of radial and circumferential stresses distribution, respectively, for the homogeneous hollow sphere ($m = 0$ in Eq.(44)) subjected to an internal pressure without considering magnetic field and thermal shock. For convenience of comparison with reference [21], the same parameters are taken: the ratio of internal radius to external radius $q = 0.5$, the dimensionless radial coordinate $\bar{r} = r/b$ and dimensionless time $\tau = V_e t/b$. The boundary conditions are expressed as $P_i(t) = H(\tau), P_o(t) = 0$. The results shown in Figs. (10) and (11) are in reasonable conformance with the Wang's et al. [21].

CONCLUSIONS

In this chapter an analytical method is developed to obtain the response of magneto-thermo-elastic stress and perturbation of the magnetic field vector for a thick-walled spherical FGM vessel. Considering the obtained results, the following conclusions are made:

1. The responses of magnetothermoelastic waves are different for different points along the radius.
2. The increasing of the magnetic field vector has a considerable effect on the frequency and the speed of the wave. The wave speed increases with increasing the magnetic field intensity.
3. The material inhomogeneity has a considerable effect on the response of the waves, as the material inhomogeneity increases, the quasi-static solution and the maximum amplitude of these waves change. Therefore, using this characteristic and varying the material in-homogeneity constant, is possible to design and optimum FGM sphere.

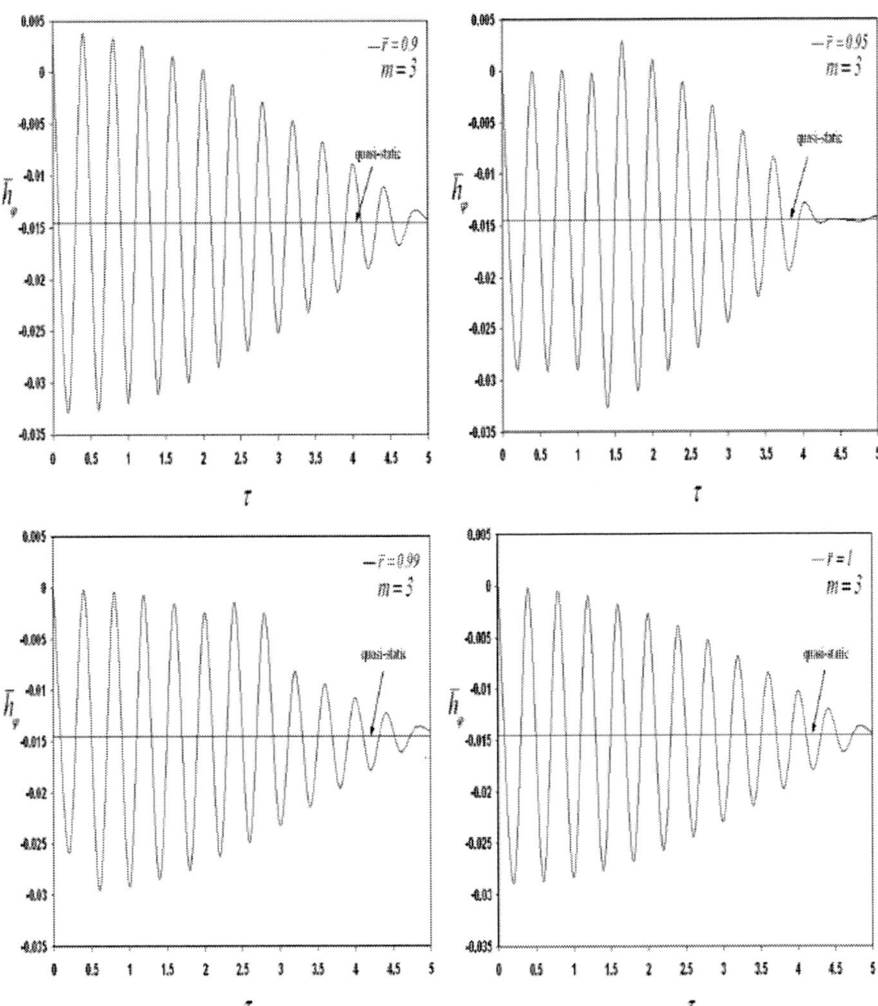

Figure 3. Response histories of perturbation of magnetic field vector in a functionally graded hollow sphere subjected to an internal pressure and thermal shock placed in a uniform circumferential magnetic field, where $\bar{r}=r/b$, $b=1m$, $\tau=V_e t/b$ and $\bar{h}_\varphi=h_\varphi/H_\varphi$.

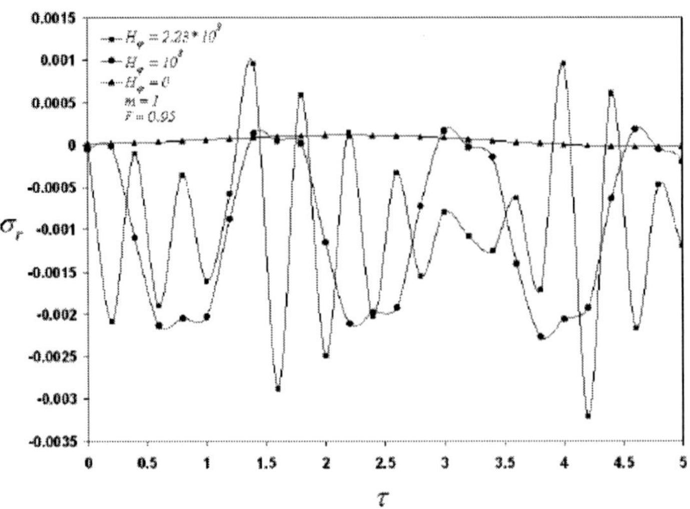

Figure 4. Response histories of radial stress in a functionally graded hollow sphere subjected to an internal pressure and thermal shock placed in a uniform circumferential magnetic field, at $\bar{r} = 0.95$ for $m = 1$ and various magnetic intensity vectors.

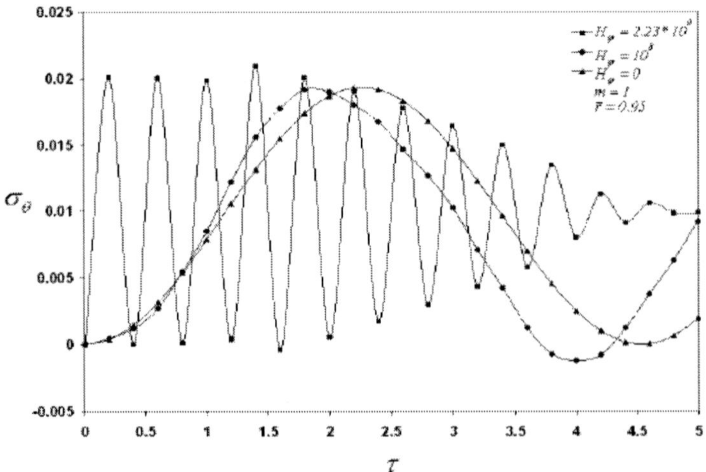

Figure 5. Response histories of circumferential stress in a functionally graded hollow sphere subjected to an internal pressure and thermal shock placed in a uniform circumferential magnetic field, at $\bar{r} = 0.95$ for $m = 1$ and various magnetic intensity vectors.

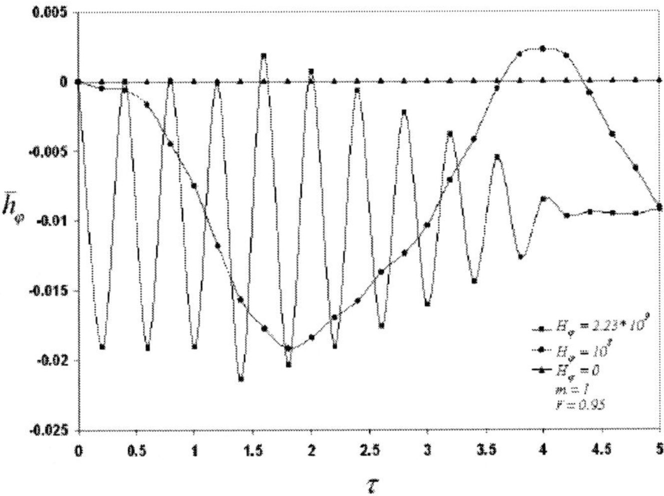

Figure 6. Response histories of perturbation of magnetic field vector in a functionally graded hollow sphere subjected to an internal pressure and thermal shock placed in a uniform circumferential magnetic field, at $\bar{r} = 0.95$ for $m = 1$ and various magnetic intensity vectors.

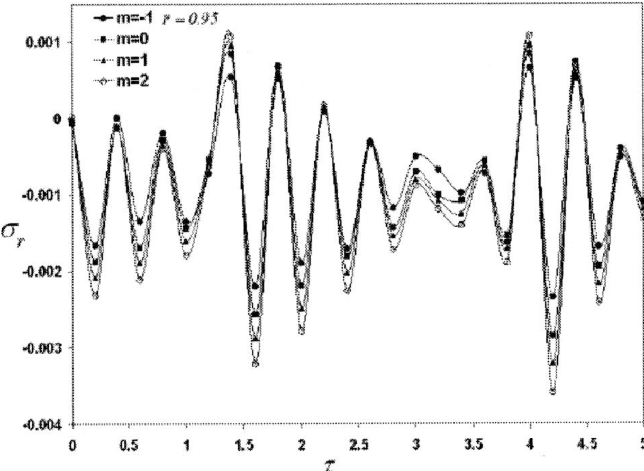

Figure 7. Response histories of radial stress in a functionally graded hollow sphere subjected to an internal pressure and thermal shock placed in a uniform circumferential magnetic field, at $\bar{r} = 0.95$ for $m = -1,0,1,2,3$.

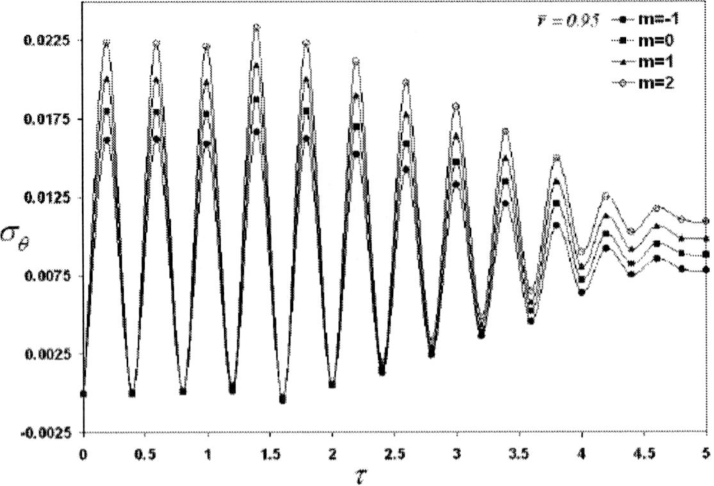

Figure 8. Response histories of circumferential stress in a functionally graded hollow sphere subjected to an internal pressure and thermal shock placed in a uniform circumferential magnetic field, at $\bar{r} = 0.95$ for $m = -1,0,1,2,3$.

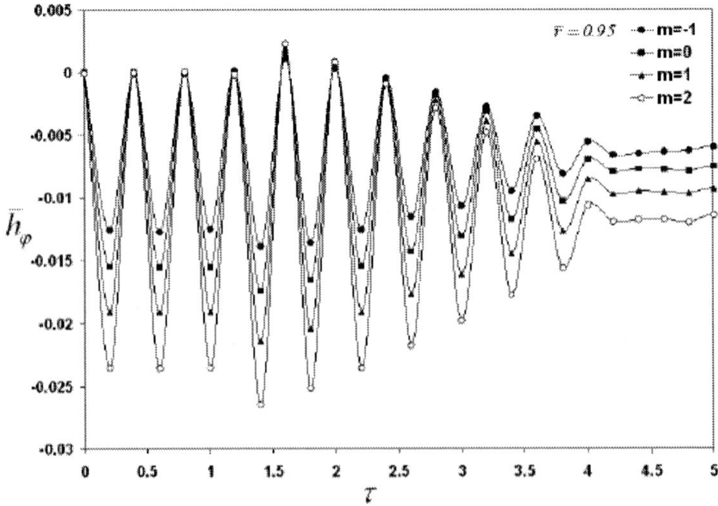

Figure 9. Response histories of perturbation of magnetic field vector in a functionally graded hollow sphere subjected to an internal pressure and thermal shock placed in a uniform circumferential magnetic field, at $\bar{r} = 0.95$ for $m = -1,0,1,2,3$.

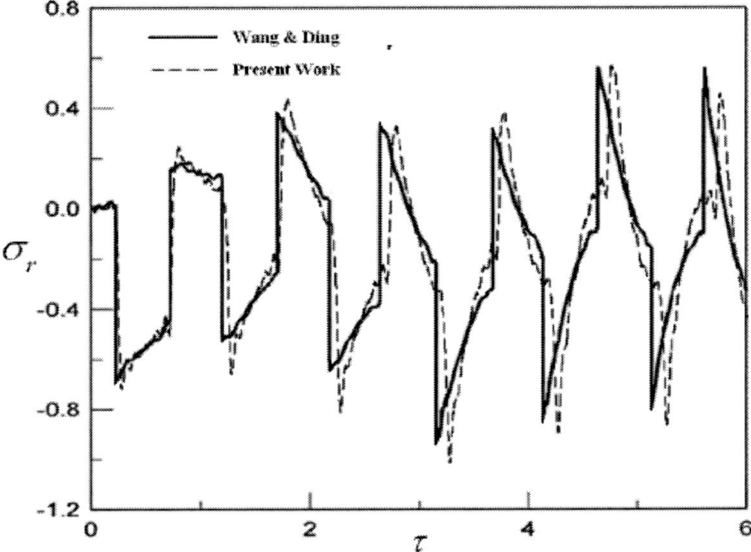

Figure 10. Response histories of radial stress in a homogeneous hollow sphere at $\bar{r} = 0.75$ without considering a magnetic field load and thermal shock.

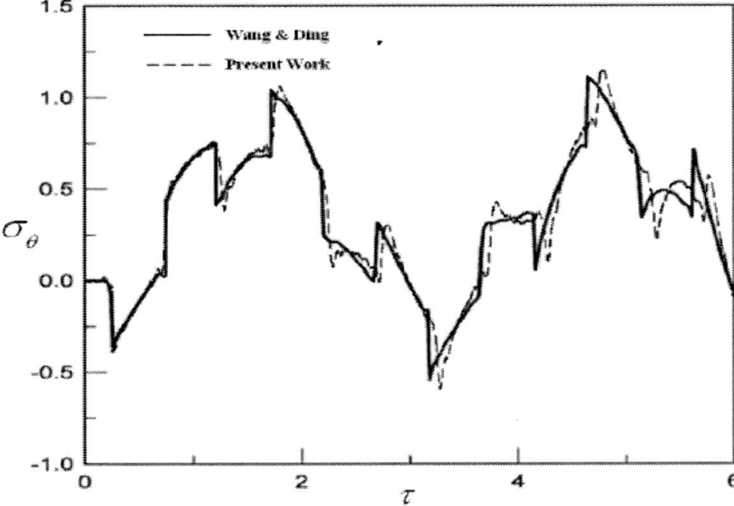

Figure 11. Response histories of circumferential stress in a homogeneous hollow sphere at $\bar{r} = 0.75$ without considering a magnetic field load and thermal shock.

1. The solution method developed above is quite accurate and is applicable for solving magneto-thermo-elastic problems of FGM with different boundary conditions and material in-homogeneity.
2. The results indicated in Figs. 10 and 11 have reasonable conformance with Wang's et al. [21]. Therefore, it is conclude that from this study, the analyses and numerical results presented in the chapter are accurate and reliable and may be used as a reference to solve other dynamic coupled problems in a FGM hollow sphere placed in a uniform magnetic field, is subjected to mechanical load and thermal shock.

REFERENCES

[1] Nayfeh A., Nemat-Nasser S., Electromagneto-thermoelastic plane waves in solids with thermal relaxation. *J. Appl. Mech. Series* E 1972;39:108–113.

[2] Choudhuri S., Electro-magneto-thermo-elastic plane waves in rotating media with thermal relaxation. *Int. J. Eng. Sci.* 1984;22:519–530.

[3] Dai H.L., Fu Y.M., Dong Z.M. Exact solution for functionally graded pressure vessels in a uniform magnetic field. *Int. J. Solids and Structures* 2006; 43:5570-5580

[4] Hosseini S.M., Akhlaghi M., Shakeri M., Transient heat conduction in functionally graded thick hollow cylinders by analytical method. *J. Heat Mass Transfer,* 2007; 43:669-675

[5] Ootao Y., Tanigawa Y., Transient piezothermoelastic analysis for a functionally graded thermo-piezoelectric hollow sphere. *J. Composite Structures,* 2007; 81:540-549

[6] Chand D., Sharma, J.N., Sud, S.P., Transient generalized magneto-thermo-elastic waves in a rotating half-space. *Int. J. Eng. Sci.* 1990; 28:547–556.

[7] Sherief H.H., Ezzat M.A., Thermal-shock problem in magneto-thermoelasticity with thermal relaxation. Int. J. Solids Struct. 1996; 33(30):4449–4459.

[8] Ezzat M.A., Generation of generalized magnetotermoelastic waves by thermal shock in a perfectly conducting half-space. *J. Thermal Stresses* 1997;20(6):617–633.

[9] Abd-Alla A.M., Abd-Alla A.N., Zeidan N.A., Transient thermal stresses in a rotation non-homogeneous cylindrically orthotropic composite tubes. *J. Appl. Math. Comput.* 1999;105:253–269.

[10] Abd-Alla A.M., El-Naggar A.M., Fahmy M.A. Magneto-thermoelastic problem in nonhomogeneous isotropic cylinder. Heat Mass Transf. 1999;39:625–629.

[11] Wang X., Lu G., Magnetothermodynamic stress and perturbation of magnetic field vector in a solid cylinder. *Journal of Thermal Stresses* 2002;25:909–926.

[12] Dai H.L., Wang X., Dynamic responses of piezoelectric hollow cylinders in an axial magnetic field. *International Journal of Solids and Structures* 2004;41:5231–5246.

[13] Wang X., Dai H.L., Magenetothermodynamic stress and perturbation of magnetic field vector in an orthotropic thermoelastic cylinder, *International journal of Engineering Science* 42(2004)539-556

[14] Dai H.L., Wang X., thermo electro elastic transient responses in piezoelectric hollow structures, *International Journal of solids and structures* 42(2005)1151-1171.

[15] Dai H.L., Fu Y.M., Magnetothermoelastic interaction in hollow structures of functionally graded material subjected to mechanical loads. *Int. J. Pressure vessels and piping* 2007;84:132-138

[16] Wang X., Dong K., Magnetothermodynamic stress and perturbation of magnetic field vector in a non-homogeneous thermoelastic cylinder. *European Journal of Mechanics A/Solids* 2006;25:98-109

[17] Dai H.L., Wang X., The dynamic response and perturbation of magnetic field vector of orthotropic cylinders under various shock loads , International *Journal of pressure vessels and piping* 83(2006)55-62

[18] Abd-Alla A.M., Fahmy M.A., El-Shahat T.M. Magneto-thermo-elastic problem of a rotating nonhomogeneous anisotropic solid cylinder. *J. Arch. Appl. Mech.*, 2007.

[19] Grag M., Rao A., Kalla S.L. On a generalized finite Hankel transform. *J. Applied Mathematics and Computation* 2007;190:705-711.

[20] Kraus J.D., Electromagnetic. USA. McGraw-Hill, Inc.; 1984.

[21] Wang, H.M., Ding, H.J., Transient responses of a magnetoelectroelastic hollow sphere for fully coupled spherically symmetric problem *European Journal of Mechanics A/Solids* 2006;25:965–980.

Chapter 6

MAGNETOTHERMOELASTIC STRESS AND PERTURBATION OF MAGNETIC FIELD VECTOR IN A FUNCTIONALLY GRADED HOLLOW SPHERE

ABSTRACT

In this chapter a closed form solution for one-dimensional magnetothermoelastic problem in a functionally graded material (FGM) hollow sphere placed in uniform magnetic and temperature fields subjected to an internal pressure is obtained using the infinitesimal theory of magnetothermoelasticity. Hyper-geometric functions are employed to solve the governing equation. The material properties through the graded direction are assumed to be nonlinear with an exponential distribution. The nonhomogeneity of the material in the radial directions is assumed to be exponential. The temperature, displacement and stress fields and the perturbation of magnetic field vector are determined and compared with those of the homogeneous case. Hence, the effect of inhomogeneity on the stresses and the perturbation of magnetic field vector distributions are demonstrated. The results of this study are applicable for designing optimum FGM hollow spheres.

NOMENCLATURE

a	inner radius of sphere
b	outer radius of sphere

u	radial displacement (m)
$\vec{U}$	displacement vector
λ, G	Lame' constants (N/m^2)
α	thermal expansion coefficient (1/ ^{0}C)
α_0	nominal thermal expansion coefficient (1/ ^{0}C)
K	thermal conductivity (W/mK)
K_0	nominal thermal conductivity (W/mK)
σ_r, σ_θ	stress components (N/m^2)
T	temperature (^{0}C)
r	radial coordinate (m)
$\vec{H}$	magnetic intensity vector
$\vec{h}$	perturbation of magnetic field vector
$\vec{J}$	electric current density vector
$\vec{e}$	perturbation of electric field vector
μ	magnetic permeability (H/m)
μ_0	nominal magnetic permeability (H/m)
f_φ	Lorentz's force per unit volume (N/m3)
E	Young's Modulus
E_0	nominal Young's Modulus
ν	Poisson's ratio

1. INTRODUCTION

Up to now, there have only been a few studies on the magneto-thermo-elastic behavior of functionally graded materials. Lutz and Zimmerman obtained the analytical solution for the stresses in spheres and cylinders made of FGMs [1, 2]. They considered thick spheres and cylinders under radial thermal loads with linear composition of the constituent materials

You and Zhang presented an accurate method to carry out elastic analysis of thick-walled spherical pressure vessels subjected to an internal pressure. They considered two kinds of pressure vessels consisting of two homogeneous layers in their inner and outer surfaces and one functionally graded layer in the middle [3]. Obata and Noda studied the thermal stresses in a functionally graded circular hollow cylinder and a hollow sphere using the perturbation method assuming one-dimensional steady-state conditions [4]. Using the infinitesimal theory of elasticity, Eslami et al. obtained closed-form solutions for stress and displacement in a functionally graded thick sphere subjected to the thermal and mechanical loads [5]. Nayebi and Abdi developed a numerical scheme to investigate the steady state behavior of thick-walled spherical and cylindrical pressure vessels under pressure and non uniform temperature field using linear kinematic hardening and Norton power law models in the plastic and creep conditions, respectively [6]. Dai et al. studied the magnetoelastic behavior of FGM cylindrical and spherical vessels subjected to an internal pressure and placed in a uniform magnetic field analytically [7]. Dai and Fu considered the magneto-thermo-elastic problem of FGM hollow structures subjected to mechanical loads. They assumed that the material properties obey the simple power-law variation through the structures' wall thickness [8].

However, exact solutions for FGM hollow spheres, with material properties obeying the exponential–power law variation subjected to an internal pressure, and placed in uniform magnetic and temperature fields remain to be found.

In this chapter, exact solutions for the stresses, displacement and perturbation of the magnetic field vector in FGM hollow spheres are obtained using the infinitesimal theory of magnetothermoelasticity. The direct method is used to solve the heat conduction and Navier equations. The FGM properties of the sphere, except Poisson's ratio, are assumed to depend exponentially on the radius. The objective of the present study is to investigate the effect of material inhomogeneity on the magneto-thermo-elastic stresses and perturbation of magnetic field vector and thereby to be able to design optimum FGM hollow spheres.

2. Heat Conduction Problem

The steady - state heat conduction equation and the boundary conditions for the one-dimensional problem in spherical coordinates for an FGM hollow sphere are given by

$$\begin{aligned} &\frac{1}{r^2}(r^2K(r)T'(r))' = 0, \\ &A_{11}T'(a) + A_{12}T(a) = f_1, \\ &A_{21}T'(b) + A_{22}T(b) = f_2, \end{aligned} \qquad \text{(1a) (1b)}$$

where, $K = K(r)$ is the thermal conductivity and A_{ij}, $i, j = 1,2$ can either designate the thermal conductivity or the heat transfer coefficient depending on the type of thermal boundary conditions employed. The constants f_1 and f_2 are known constants on the inner and outer radius, respectively. It is assumed that the non-homogeneous thermal conductivity $K(r)$ is an exponential function of radius as

$$K = K_0 e^{k(\frac{r}{a})^n}, \tag{2}$$

where, k and n are material parameter. Using Eq. (2) for the thermal conductivity, the heat conduction equation becomes

$$\frac{1}{r^2}(r^2 K_0 e^{k(\frac{r}{a})^n} T'(r))' = 0 \tag{3}$$

Integrating Eq. (3) twice yields

$$T(r) = A_1 \int_a^r \frac{e^{-k(\frac{r}{a})^n}}{r^2} dr + A_2. \tag{4}$$

Applying the boundary conditions (1b) results in the following relations for the coefficients A_1 and A_2

$$A_1 = \frac{A_{22} f_1 - A_{12} f_2}{A_{11} A_{22} \frac{e^{-k}}{a^2} - A_{12} A_{21} \frac{e^{-k(\frac{b}{a})^n}}{b^2} - A_{12} A_{22} \int_a^b \frac{e^{-k(\frac{r}{a})^n}}{r^2} dr}, \tag{5}$$

$$A_2 = \frac{-f_1(A_{21}\dfrac{a^2 e^{-k(\frac{b}{a})^n}}{b^2} + A_{22}a^2\int_a^b \dfrac{e^{-k(\frac{r}{a})^n}}{r^2}dr) + A_{11}e^{-k}f_2}{A_{11}A_{22}e^{-k} - A_{12}A_{21}\dfrac{a^2 e^{-k(\frac{b}{a})^n}}{b^2} - A_{12}A_{22}a^2\int_a^b \dfrac{e^{-k(\frac{r}{a})^n}}{r^2}dr}. \tag{6}$$

3. Basic Formulations and Solutions

Consider a thick hollow sphere of inner radius a and outer radius b inside with perfect conductivity placed in a uniform magnetic field $\vec{H}(0,0,H_\phi)$ made of FGM. Let u be the displacement component in the radial direction. Then, the strain–displacement relations are

$$\varepsilon_{rr} = \frac{du}{dr}, \qquad \varepsilon_{\theta\theta} = \frac{u}{r}. \tag{7}$$

The stress-strain relations are

$$\begin{aligned}\sigma_{rr} &= \lambda e + 2G\varepsilon_{rr} - (3\lambda + 2G)\alpha T(r),\\ \sigma_{\theta\theta} &= \lambda e + 2G\varepsilon_{\theta\theta} - (3\lambda + 2G)\alpha T(r),\end{aligned} \tag{8}$$

where, $\sigma_{ij}, \varepsilon_{ij}$ ($i,j = r,\theta$) are the stress and strain tensors, e is the first invariants of strain tensor, $T(r)$ is the temperature distribution determined from the Eq.(4), α is the coefficient of thermal expansion, and λ and G are Lame′ coefficients which are given by

$$\lambda = \frac{\nu E}{(1+\nu)(1-2\nu)}, \qquad G = \frac{E}{2(1+\nu)}, \tag{9}$$

where, E is the elasticity modulus and ν is the Poisson's ratio. Assuming that the magnetic permeability, μ, at the outer surface of the FGM sphere to be equal to the magnetic permeability of the medium around it, the medium to be non-ferromagnetic and non-ferroelectric, and omitting displacement electric currents, the governing electrodynamic Maxwell equations for a perfectly conducting elastic body can be written as [9, 10]

$$\vec{J} = \nabla \times \vec{h}, \qquad \nabla \times \vec{e} = -\mu \frac{\partial \vec{h}}{\partial t}, \qquad div\vec{h} = 0,$$
$$\vec{e} = -\mu\left(\frac{\partial \vec{U}}{\partial t} \times \vec{H}\right), \qquad \vec{h} = Curl\left(\vec{U} \times \vec{H}\right) \tag{10}$$

Applying an initial magnetic field vector $\vec{H}(0,0,H_{\phi})$ in the spherical coordinates (r,θ,ϕ) to Eq. (10), yields

$$\vec{U} = (u,0,0), \quad \vec{e} = -\mu\left(0, H_{\varphi}\frac{\partial u}{\partial t}, 0\right),$$
$$\vec{h} = (0,0,h_{\varphi}), \vec{J} = \left(0, -\frac{\partial h_{\varphi}}{\partial r}, 0\right), \ h_{\varphi} = -H_{\varphi}\left(\frac{\partial u}{\partial r} + \frac{2u}{r}\right). \tag{11}$$

The equilibrium equation of the FGM hollow sphere in the absence of body forces is expressed as

$$\frac{\partial \sigma_{rr}}{\partial r} + \frac{2}{r}(\sigma_{rr} - \sigma_{\theta\theta}) + f_{\varphi} = 0. \tag{12}$$

where f_{ϕ} is the Lorentz's force [9,10] which may be written as

$$f_{\varphi} = \mu(r)\left(\vec{J} \times \vec{H}\right) = \mu(r)H^2{}_{\varphi}\frac{\partial}{\partial r}\left(\frac{\partial u}{\partial r} + \frac{2u}{r}\right). \tag{13}$$

To obtain the equilibrium equation in terms of the displacement for the FGM sphere, the functional relationship of the material properties must be known. The sphere's material is assumed to be described with an exponential function of the radial direction as

$$E = E_0 e^{k(\frac{r}{a})^n}, \ \alpha = \alpha_0 e^{k(\frac{r}{a})^n}, \mu = \mu_0 e^{k(\frac{r}{a})^n}. \tag{14}$$

We further assume that Poisson's ratio is constant. Using relations (7)-(14), the Navier equation in term of the displacement is written as

$$\frac{d^2u}{dr^2} + (\frac{2}{r} + \frac{A\,kn}{(A+\mu_0 H_\varphi{}^2)a^n} r^{n-1})\frac{du}{dr} + (\frac{knN}{a^n} r^{n-2} + \frac{M}{r^2})u =$$
$$\frac{2knC\alpha_0}{a^n(A+\mu_0 H_\varphi{}^2)} r^{n-1} e^{k(\frac{r}{a})^n} T(r) + \frac{A_1 C\alpha_0}{r^2(A+\mu_0 H_\varphi{}^2)}. \tag{15}$$

Hence, the non-homogeneous term F is found to be equal to

$$F(r) = \frac{2knC\alpha_0}{a^n(A+\mu_0 H_\varphi{}^2)} r^{n-1} e^{k(\frac{r}{a})^n} T(r) + \frac{A_1 C\alpha_0}{r^2(A+\mu_0 H_\varphi{}^2)}, \tag{16}$$

where, M and N are given by

$$M = 2\left(\frac{B(1-1/\nu) - \mu_0 H_\theta{}^2}{A+\mu_0 H_\varphi{}^2}\right), \ N = 2\left(\frac{B}{A+\mu_0 H_\varphi{}^2}\right).$$

The coefficients A, B, and C are

$$A = \frac{(1-\nu)E_0}{(1+\nu)(1-2\nu)},$$
$$B = \frac{\nu E_0}{(1+\nu)(1-2\nu)},$$
$$C = \frac{E_0}{1-2\nu}.$$

The general solution of Eq. (15) can be written as

$$u(r) = C_1 P(r) + C_2 Q(r) + R(r), \tag{17}$$

where, C_i is an arbitrary integration constant, P and Q are homogeneous solutions, and R is the particular solution. Using the method of variation of parameters, the particular solution is written as

$$R(r) = -P(r)\int_0^r \frac{Q(\zeta)F(\zeta)}{W(\zeta)}d\zeta + Q(r)\int_0^r \frac{P(\zeta)F(\zeta)}{W(\zeta)}d\zeta \quad , \tag{18}$$

where,

$$W(r) = P(r)\frac{dQ}{dr} - Q(r)\frac{dP}{dr}. \tag{19}$$

The non-homogeneous differential equation, Eq.(15), is of confluent hyper-geometric type which may be solved by introducing the transformations

$$u(r) = r^{\frac{-1+\sqrt{1-4M}}{2}} y$$

and

$$x = -\frac{A\,k\,(\frac{r}{a})^n}{A + \mu_0 H_\varphi^{\ 2}}.$$

Using these transformations, the homogeneous part is transformed into

$$x\frac{d^2y}{dx^2} + \frac{dy}{dx}[\beta - x] - \chi y = 0 \quad , \tag{20}$$

where, χ and β are given by

$$\chi = \frac{2N + A(-1+\sqrt{1-4M})}{2An},$$
$$\beta = \frac{n+\sqrt{1-4M}}{n}. \tag{21}$$

Equation (20) is a standard confluent hyper-geometric differential equation and its solution is written as [11]

$$y(x) = \hat{C}_1 F_c(\chi;\beta;x) + \hat{C}_2 x^{1-\beta} F_c(1+\chi-\beta;2-\beta;x), \tag{22}$$

where, $F_c(\chi;\beta;x)$ is the confluent hyper-geometric function defined by

$$F_c(\chi;\beta;x) = 1 + \frac{\chi}{\beta.1!}x + \frac{\chi(\chi+1)}{\beta(\beta+1).2!}x^2 + \frac{\chi(\chi+1)(\chi+2)}{\beta(\beta+1)(\beta+2).3!}x^3 + ... \tag{23}$$

Finally, using $u(r) = r^{\frac{-1+\sqrt{1-4M}}{2}} y$ and $x = -\frac{Ak(\frac{r}{a})^n}{A+\mu_0 H_\varphi^{\,2}}$, the following homogeneous solutions are obtained

$$P(r) = r^{\frac{-1+\sqrt{1-4M}}{2}} F_c(\chi;\beta;-\frac{Ak(\frac{r}{a})^n}{A+\mu_0 H_\varphi^{\,2}}),$$
$$Q(r) = r^{\frac{-1+\sqrt{1-4M}}{2}+n(1-\beta)} F_c(1+\chi-\beta;2-\beta;-\frac{Ak(\frac{r}{a})^n}{A+\mu_0 H_\varphi^{\,2}}). \tag{24}$$

By the virtue of Eqs. (18) and (24), the expression for the stresses and perturbation of magnetic field vector in the FGM hollow sphere are derived as follows:

$$\sigma_r = C_1 V(r) + C_2 H(r) + S(r)\,, \tag{25}$$

$$\sigma_\theta = C_1 \overline{V}(r) + C_2 \overline{H}(r) + \overline{S}(r)\,, \tag{26}$$

$$\begin{aligned} h_\phi = -H_\phi [C_1 (\frac{dP(r)}{dr} + 2r^{-1}P(r)) \\ + C_2 (\frac{dQ(r)}{dr} + 2r^{-1}Q(r)) + \frac{dR(r)}{dr} + 2r^{-1}R(r) \end{aligned} \tag{27}$$

where,

$$\begin{aligned} V(r) &= (\lambda + 2G)\frac{dP}{dr} + \frac{2\lambda P(r)}{r}, \\ H(r) &= (\lambda + 2G)\frac{dQ}{dr} + \frac{2\lambda Q(r)}{r}, \\ S(r) &= (\lambda + 2G)\frac{dR}{dr} + \frac{2\lambda R(r)}{r} - (3\lambda + 2G)\alpha T(r), \end{aligned} \tag{28}$$

$$\begin{aligned} \overline{V}(r) &= \lambda\frac{dP}{dr} + \frac{2(\lambda + G)P(r)}{r}, \\ \overline{H}(r) &= \lambda\frac{dQ}{dr} + \frac{2(\lambda + G)Q(r)}{r}, \\ \overline{S}(r) &= \lambda\frac{dR}{dr} + \frac{2(\lambda + G)R(r)}{r} - (3\lambda + 2G)\alpha T(r). \end{aligned} \tag{29}$$

To determine the constants C_1 and C_2, the boundary conditions for stresses need to be applied. Consider the mechanical boundary conditions on the inside and outside radii as

$$\sigma_{rr}(a) = -P_i, \qquad \sigma_{rr}(b) = -P_o\,. \tag{30}$$

Substituting the above boundary conditions into Eq. (25) yields the following constants of integration

$$C_1 = \frac{P_i H(b) - P_o H(a) + S(a)H(b) - S(b)H(a)}{V(b)H(a) - V(a)H(b)},$$

$$C_2 = \frac{P_o \dfrac{V(a)(H(a)+H(b)) - V(b)H(a)}{H(a)} - P_i \dfrac{V(a)H(b)}{H(a)} + S(b)V(a) - S(a)V(b)}{V(b)H(a) - V(a)H(b)}. \qquad (31)$$

4. Numerical Results and Discussion

To conduct the numerical calculations, the material constants $E_0 = 200\,Gp$, $\alpha_0 = 1.2\times10^{-6}\,(1/K)$, $\mu_0 = 4\pi\times10^{-7}(H/m)$, $\nu = 0.3$ are used for the FGM hollow sphere. Magnetic intensity is taken as $H_\phi = 2.23\times10^{9}\,(A/m)$. The temperature, displacements, magneto-thermo-elastic stresses and the perturbation of the magnetic field vector for the FGM hollow spheres subjected to mechanical loads are normalized to demonstrate the effect of inhomogeneity.

Test case 1. As the first example, consider a thick hollow sphere of inner radius $a = 0.5$ m and outer radius $b = 1\,\mathrm{m}$. The boundary conditions for temperature are taken as $T(a) = 10^\circ C$ and $T(b) = 0^\circ C$ .The hollow sphere is assumed to be under internal pressure of $200 Mpa$ and a zero external pressure, i.e. $\sigma_{rr}(a) = -200\ Mpa$ and $\sigma_{rr}(b) = 0\,Mpa$.

The temperature profile, radial displacement, radial stresses, hoop stresses, effective stresses and perturbation of magnetic field vector along the radial direction for $n = 1.2$ and different values of k are shown in Figs. 1 through 6.

Figure 1 shows the variation of temperature for different values of material parameter k . It can be seen from this figure that with increasing the parameter k , the temperature decreases. Figure 2 shows that the radial displacement decreases with increasing k . Moreover, it tends to constant values throughout the wall thickness for high values of k .

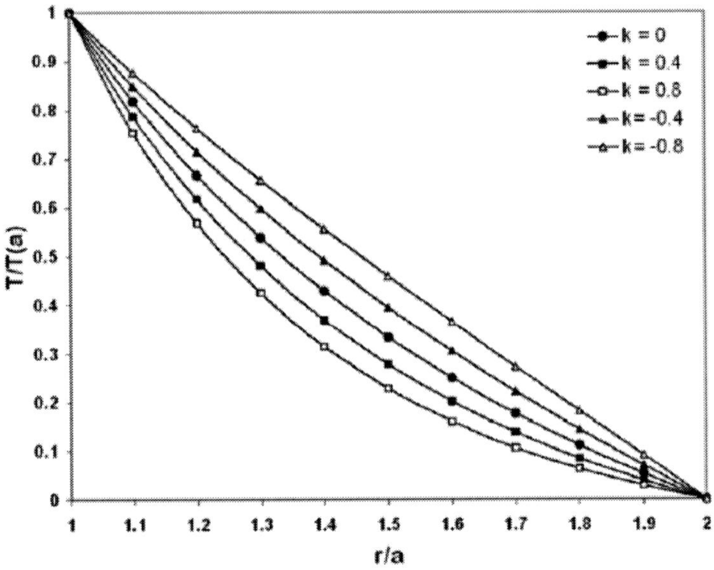

Figure 1. Radial temperature distribution for $n = 1.2$.

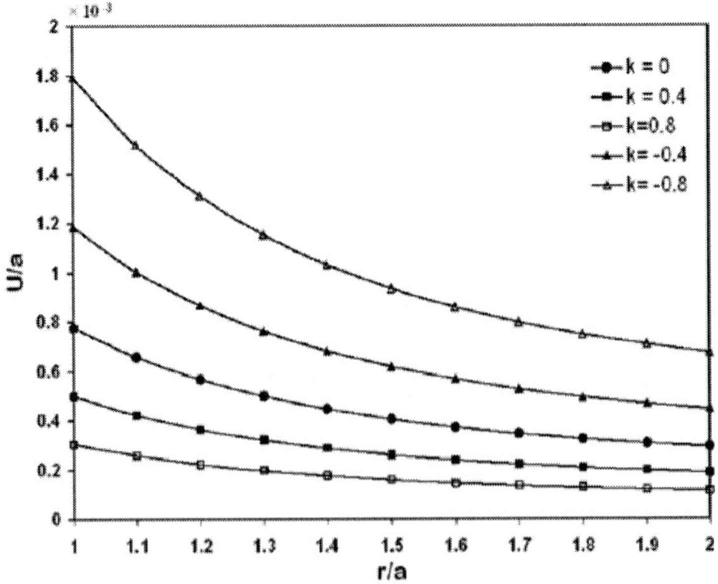

Figure 2. Radial displacement distribution for $n = 1.2$.

Figure 3, depicts the distribution of radial stress. It is seen from Figure 2 that the radial stresses at the internal and external surfaces of the FGM hollow sphere satisfy the given boundary conditions. Moreover the curves in this figure show a nearly linear distribution of stresses with increasing parameter k.

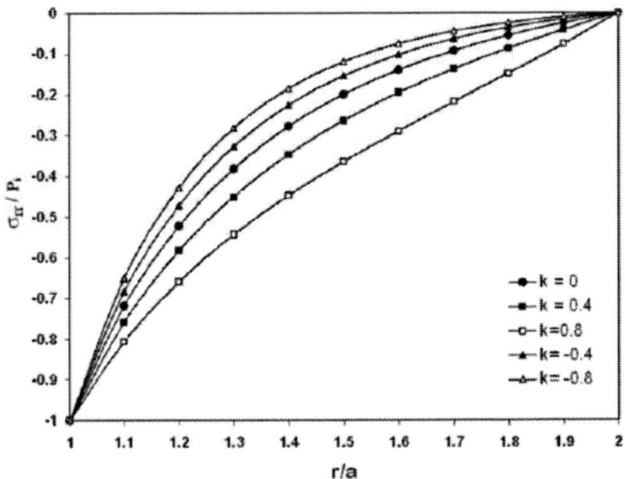

Figure 3. Radial stress distribution for $n = 1.2$.

It is seen from Figure 4 that as the parameter k increases, the circumferential stresses decrease near the inner surface of FGM hollow sphere while opposite results are observed in other points.

Figure 5 shows the distribution of the effective stress $\sigma_e = \sqrt{2}|\sigma_{rr} - \sigma_{\theta\theta}|$ [5]. As can be seen from this figure, the variation of the effective stresses is similar to that of Figure 4. The distributions of the perturbation of magnetic field vector are plotted in Figure 6. It is seen that with increasing k, the magnitude of perturbation of magnetic field vector becomes small.

Test case 2. For the second example $\sigma_{rr}(a) = 0\, Mpa$ and $\sigma_{rr}(b) = 0\, Mpa$ are considered. Figures 7 through 11 show the distribution of radial displacement, radial stress, circumferential stress, effective stress and perturbation of magnetic field vector in the FGM hollow sphere for this case.

The radial displacement curves in Figure 7 show a trend contrary to those of the curves in Figure2. It is seen from Figure 8 that the radial stress at the internal and external surfaces of the FGM hollow sphere satisfy the given boundary conditions and as the parameter k increases the magnitude of radial stress becomes large.

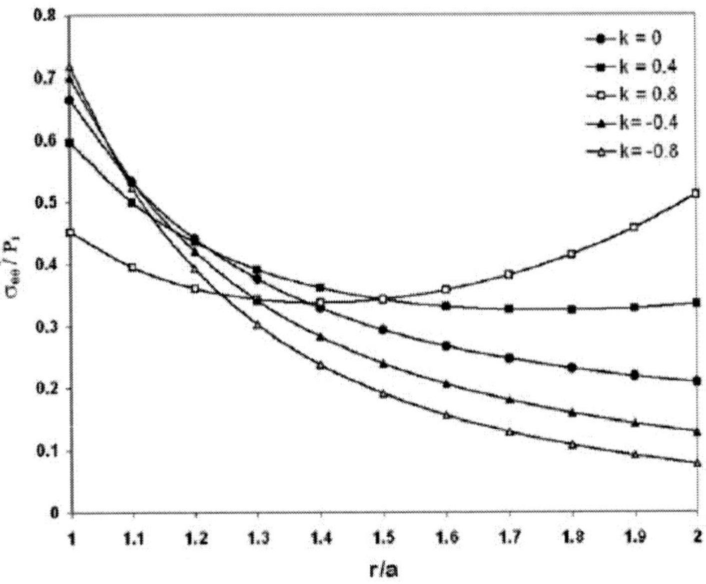

Figure 4. Circumferential stress distribution for $n = 1.2$.

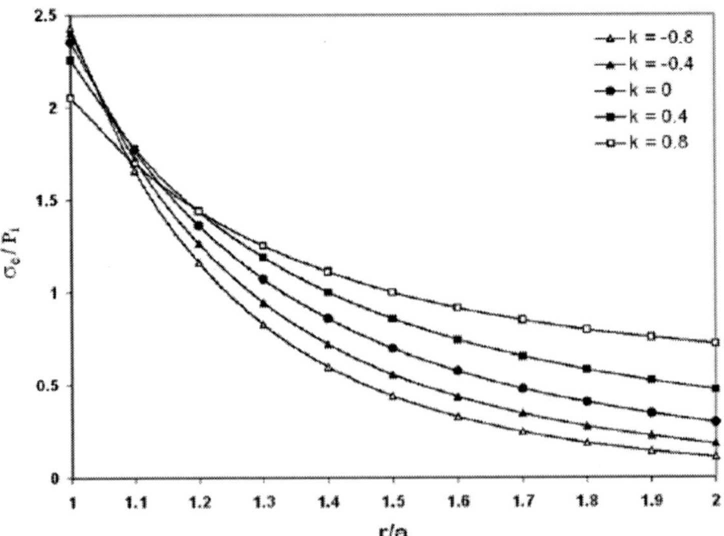

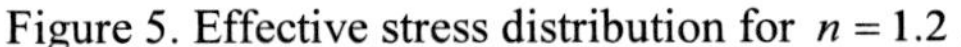

Figure 5. Effective stress distribution for $n = 1.2$.

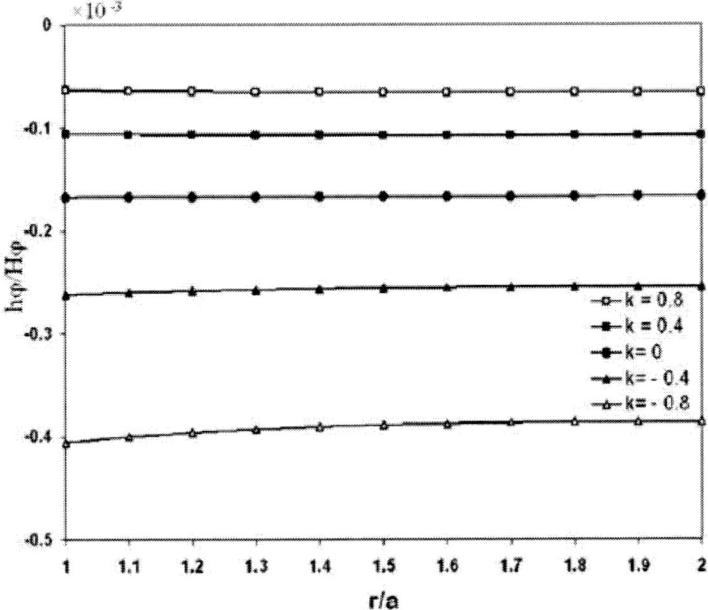

Figure 6. The perturbation of the magnetic field vector distribution for $n = 1.2$.

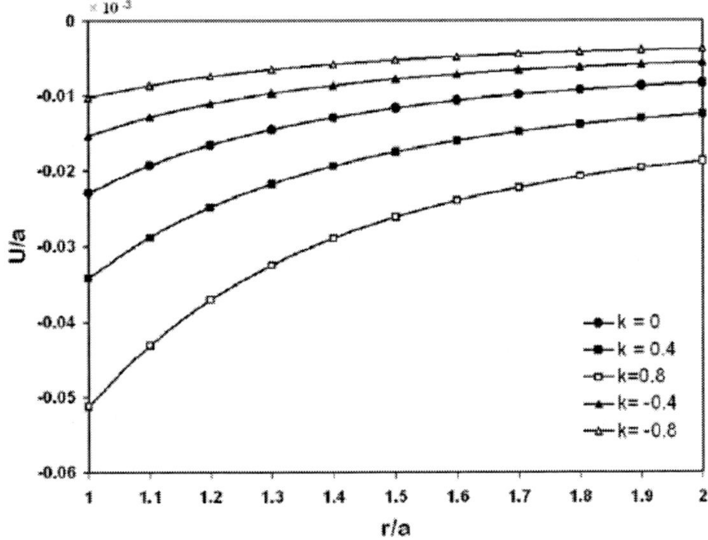

Figure 7. Radial displacement distribution.

Figure 10 shows the distribution of the effective stress. It is seen from this figure that the effective stress increases with increasing the parameter k. Considering the results of Figure 11, it is observed that the parameter k has a significant effect on the perturbation of magnetic field vector distribution.

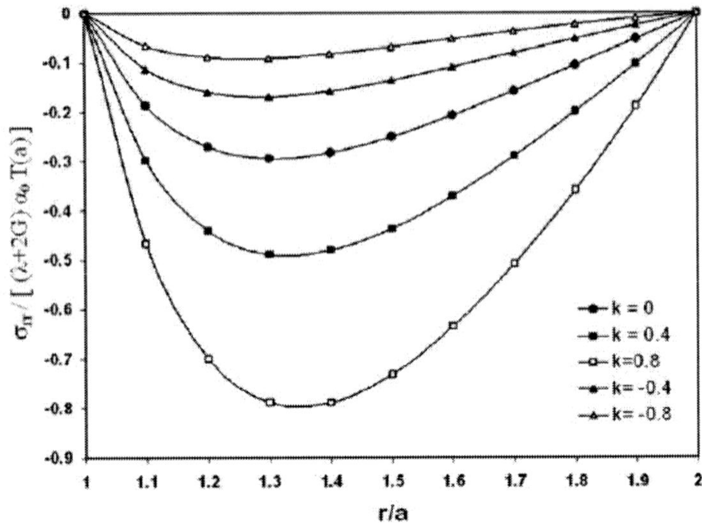

Figure 8. Radial stress distribution.

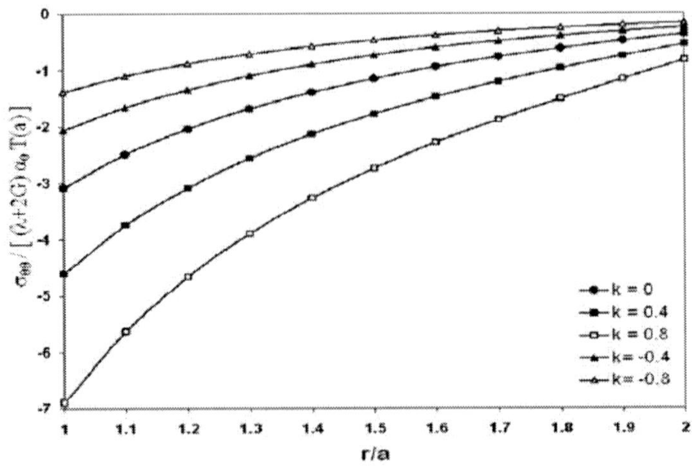

Figure 9. Circumferential stress distribution.

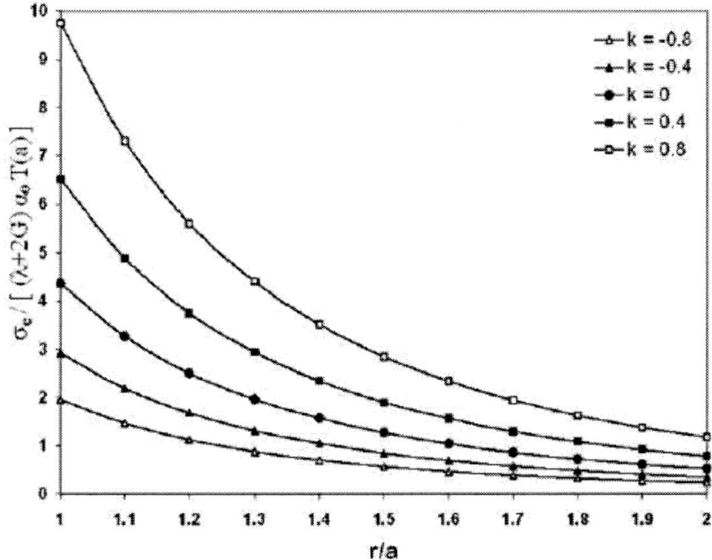

Figure 10. Effective stress distribution.

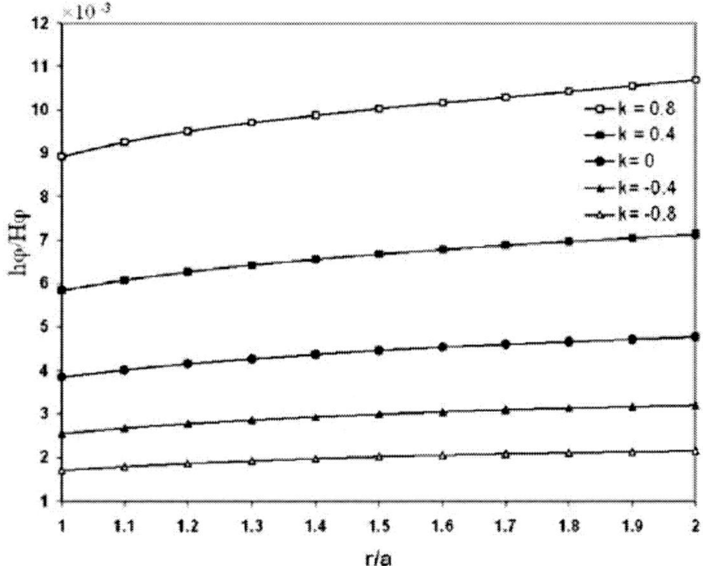

Figure 11. The perturbation of the magnetic field vector distribution.

CONCLUSIONS

Analytical solutions for FGM hollow sphere placed in a uniform magnetic field and temperature field subjected to internal pressure are obtained. The modeling presented in this chapter is based on the direct method of solution of the heat conduction and Navier equations. The exact solution is found by solving a hyper-geometric equation. The sphere's material is assumed to be graded along the r- direction. The material inhomogeneities vary exponentially along the radial direction.

Numerical results show the significant influence of material inhomogeneity on the magnetothermoelastic stresses and perturbation of magnetic field vector. Using the results of this study, it is possible to find optimum values for exponential-power law indices k and n such that the circumferential stress is almost constant with minimal variation along the radial direction.

REFERENCES

[1] Lutz MP., Zimmerman RW. Thermal stresses and effective thermal expansion coefficient of a functionally graded sphere.*J. Thermal Stress* 1996;19:39-54.

[2] Zimmerman RW., Lutz MP. Thermal stresses and effective thermal expansion in a uniformly heated functionally graded cylinder. *J. Thermal Stress* 1999;22:177-88.

[3] You L.H., Zhang J.J., You X.Y. Elastic analysis of internally pressurized thick-walled spherical pressure vessels of functionally graded materials.*Int. J. Pressure vessels and piping* 2005;82:347-354 .

[4] Obata Y, Noda N. Steady thermal stresses in a hollow circular cylinder and a hollow sphere of a functionally gradient material. *J. Thermal Stresses* 1994;17:471-87.

[5] Eslami M.R., Babaei M.H., Poultangari R. Thermal and mechanical stresses in a functionally graded thick sphere.*Int. J. Pressure vessels and piping* 2005;82:522-527.

[6] Nayebi A, El Abdi R. Cyclic plastic and creep behaviour of pressure vessels under thermomechanical loading. *Comput. Mater. Sci.* 2002;25:285-96

[7] Dai H.L., Fu Y.M., Dong Z.M. Exact solution for functionally graded pressure vessels in a uniform magnetic field. *Int. J. Solids and Structures* 2006;43:5570-5580

[8] Dai H.L., Fu Y.M., Magnetothermoelastic interaction in hollow structures of functionally graded material subjected to mechanical loads. *Int. J. Pressure vessels and piping* 2007;84:132-138

[9] Kraus J.D., Electromagnetic. USA. McGraw-Hill, Inc.; 1984.

[10] Dai H.L., Wang X. Dynamic responses of piezoelectric hollow cylinders in an axial magnetic field. *Int. J. solids and structures* 2004;41:5231-5246

[11] Seaborn J.B. Hypergeometric functions and their applications. New york. Springer-Verlag, Inc.;1991

INDEX